Hungry Dreams

Food Systems and Agrarian Change

Edited by Frederick H. Buttel, Billie R. DeWalt,
and Per Pinstrup-Andersen

A complete list of titles in the series appears at the end of this book.

HUNGRY DREAMS

*The Failure of Food Policy
in Revolutionary Nicaragua,
1979–1990*

Brizio N. Biondi-Morra

Cornell University Press
ITHACA AND LONDON

Copyright © 1993 by Cornell University

All rights reserved. Except for brief quotations in a review, this book, or parts thereof, must not be reproduced in any form without permission in writing from the publisher. For information, address Cornell University Press, Sage House, 512 East State Street, Ithaca, New York 14850.

First published 1993 by Cornell University Press.

Library of Congress Cataloging-in-Publication Data
Biondi-Morra, Brizio N.
  Hungry dreams : the failure of food policy in revolutionary Nicaragua, 1979–1990 / Brizio N. Biondi-Morra.
    p. cm. — (Food systems and agrarian change)
  Includes bibliographical references and index.
  ISBN 0-8014-2663-4
  1. Food supply—Government policy—Nicaragua. 2. Agriculture and state—Nicaragua. I. Title. II. Series.
HD9014.N52B56 1993
363.8'097285—dc20   92-56782

Printed in the United States of America

  The paper in this book meets the minimum requirements
of the American National Standard for Information Sciences—
Permanence of Paper for Printed Library Materials, ANSI Z39.48-1984.

Contents

	Tables and Figures	vii
	Acknowledgments	ix
	Bibliographical Note	xi
	List of Abbreviations	xiii
1	Introduction	1
2	Food Policy Formulation: Objectives and Strategy	14
3	Foreign Exchange	56
4	Food Prices	99
5	Wages	138
6	Credit	173
7	Palliatives	195
8	Outcomes and Conclusions	200
	Appendix Major State-owned Agribusiness Enterprises, June 1983	207
	Index	211

Tables and Figures

TABLES

1	Importance of the food system in the national economy	18
2	Distribution of rural income, 1971	28
3	Daily per capita consumption of calories, proteins, and fats by income stratum, 1970	29
4	Projected food demand	33
5	Production targets for reaching food self-sufficiency in the year 2000	34
6	Export targets for the agricultural sector	36
7	Estimated structure of land tenure in 1990 and 2000	43
8	Estimated share of output value by type of land tenure	45
9	Basic economic indicators, 1980–1985	49
10	Farm output, 1977–1985	50
11	Farm exports, 1977–1985	53
12	State Cotton Farm: adjusted profit (loss), 1981–1985	91
13	Idle capacity of food-processing plants, 1983	104
14	Beef trade balance	115
15	Exports and imports of meat, poultry, and dairy products, 1977–1985	116
16	Prices of milk and beef, 1979–1986	119

17	Purchasing power of one pound of beef, 1981 vs. 1986	128
18	Index of official and black market consumer prices of food	129
19	Index of official prices to farm producers, 1979–1986	134
20	Economically active population in the farm sector	139
21	Estimated distribution of land ownership	140
22	Basic grains production, imports, and exports, 1977–1985	153
23	State Rice Farm: nominal vs. real wages of workers	161
24	Cost of the basic consumer basket at official and market prices	161
25	Cost of the basic food basket	162
26	State Rice Farm Complex A: labor input and rice output	167
27	Share of credit by economic sector	175
28	Credit given to state-owned enterprises (APP) and estimated amounts in default, 1980–1985	176
29	State Sugar Mill: income statements	184
30	State Sugar Mill: claims on assets	186
31	Actual calculations of the wholesale price of sugar, 1985	188
32	Interest rate structure and inflation	190

FIGURES

1	Main agricultural regions of Nicaragua	20
2	State participation in the Nicaraguan food system, 1983	44
3	The structure of the Nicaraguan cotton commodity system, 1983	72
4	The structure of the Nicaraguan milk/beef commodity system, 1983	110
5	The structure of the Nicaraguan rice commodity system, 1983	155
6	The State Rice Farm organization chart	159
7	The structure of the Nicaraguan sugar commodity system, 1983	181

Acknowledgments

During the four years I spent in Nicaragua working on this book, I always felt I was witnessing a page of history whose ultimate explanation could be traced to poverty and hunger. The search for a solution to these ills is precisely what food policy entails. Within its scope and limitations, this pursuit is also the goal of my book. In my search, I became indebted to many individuals.

James Austin, Ray Goldberg, and Peter Timmer of Harvard University's Graduate School of Business Administration deserve special mention. Without their advice and assistance, this book would never have come to fruition. Helpful comments on drafts of the manuscript were provided by Forrest Colburn of Princeton University, Mario de Franco of the Harvard Institute for International Development, and Renata Villers of Columbia University.

In Nicaragua, continuous institutional support was provided by the Central American Institute of Business Administration (INCAE). For this support, I thank John Ickis, Francisco Leguizamón, and Marc Lindenberg. I also thank Alan Hoffman, a member of the INCAE faculty, for his assistance with editing and producing this book. I am grateful to the Ford Foundation, and particularly to William Carmichael, Stephen Cox, and David Winder, for the financial support of the INCAE/Ministry of Agriculture/Central Bank research project that I directed. Without the field exposure made possible by the Ford Foundation, this book could not have even been contemplated.

Numerous Nicaraguans contributed to the research and analysis that made this book possible. Although not all of them are mentioned in the book, all have my gratitude. Heartfelt thanks go to the dozens of state managers and administrators with whom I shared frustration, toil, and hope.

B. N. B.-M.

Bibliographical Note

The sources cited in this work are principally interviews I conducted with the personnel of the Ministry of Agriculture and Land Reform (MIDINRA) and internal working documents of the ministry. My work with the ministry provided me with the opportunity to converse freely and candidly with officials. Some interviews I taped, others I recorded with handwritten notes. I also had unrestricted access to accounting data, reports, working papers, and other documents. Since these documents were for internal use, the information presented was free of the kind of bias that might have been introduced if the documents were prepared for the general public. The translations of the quoted material are mine.

The MIDINRA documents I cite were generally only mimeographed copies. The number of copies printed varied, depending on the perceived need, but was generally small. Fewer than ten copies existed of some of the documents cited. Such documents usually circulated only among the minister, vice-ministers, and other ranking government officials. MIDINRA's central office in Managua did not have means of safekeeping documents. Upon the transfer of government in 1990, the office was largely bereft of records of what had happened the previous decade. Presumably, many of MIDINRA's documents have been lost or destroyed.

Abbreviations

ANAR	Association of Rice Producers
APP	Area of People's Property
ATC	Association of Rural Workers
BAMER	Banco de America
BANANIC	Nicaraguan Banana Enterprise
BANIC	Nicaraguan Bank
BCN	Central Bank of Nicaragua
BCP	Popular Credit Bank
BINMO	Banco Inmobilario
BND	National Development Bank
CACM	Central American Common Market
CAR	rural state store
CAS	Sandinista Agricultural Cooperatives
CCS	Credit and Services Cooperatives
CDD	Certificate of Exchange Availability
CIERA	Research Center for Economics and Land Reform
CORFIN	Financial Corporation of Nicaragua
CST	Sandinista Workers Center
DOGE	Entrepreneurial Organization and Administrative Division (of MIDINRA)
EEC	European Economic Community
ENABAS	National Supply Company
ENAL	Nicaraguan Cotton Enterprise
ENAMARA	National Enterprise of Land Reform Slaughterhouses
ENAZUCAR	Nicaraguan State Sugar Enterprise
ENCAFE	Nicaraguan Coffee Enterprise
ENCAR	Beef Enterprise
ENIA	Nicaraguan Enterprise of Farm Inputs
ENILAC	Enterprise of the Milk Industry

FAGANIC	Nicaraguan Federation of Cattle Rancher Associations
FAO	Food and Agriculture Organization (of the United Nations)
FNI	National Investment Fund
FONDILAC	association of private milk producers and processors
FSLN	Sandinista Front for National Liberation
GCD	General Cattle Division (of MIDINRA)
GED	General Economic Division (of MIDINRA)
INAP	Nicaraguan Institute of Public Administration
INCAE	Central American Institute of Business Administration
INE	Nicaraguan Energy Institute
INEC	Nicaraguan Institute of Statistics and Census
INFONAC	Nicaraguan Institute of National Development
INRA	Nicaraguan Institute of Agrarian Reform
JGRN	Junta of the Government for National Reconstruction
MICE	Ministry of Foreign Trade
MICOIN	Ministry of Internal Trade
MIDINRA	Ministry of Agriculture and Land Reform
MIFIN	Ministry of Finance
MIND	Ministry of Industry
MIPLAN	Ministry of Planning
MISAL	Ministry of Health
MITRAB	Ministry of Labor
NPC	National Planning Council
PAN	National Food Program
PROAGRO	National Enterprise of Farm Products
SFN	National Financial System
SNOTS	National Organizing System of Labor and Wages
SPP	Secretariat of Planning and Budget
SUCA	System of Uniform Administrative Control (of MIDINRA)
UNWFP	United Nations World Food Program

Hungry Dreams

1

Introduction

> The critical question that we should address is: How is it that the government, with all the policy instruments it has at its disposal, is unable to reach its food policy goals?
> —Government food policy analyst, Nicaragua, 1987

Upon their ascension to power in 1979, the Sandinista Front for National Liberation (FSLN) recognized that Nicaragua's economy was fundamentally agrarian. The Sandinistas decided accordingly that their development strategy would focus on the agrarian sector. But Nicaragua's agriculture had to be restructured even as it continued to generate needed employment, food, products for local industry, and foreign exchange. Under Anastasio Somoza's dictatorship, a rapacious minority had benefited from dramatic growth in the cultivation and export of cotton, coffee, sugar, and cattle, while the vast majority of Nicaraguans lived in abject misery. Rural Nicaragua generated the country's wealth, but the rural majority was poor.[1]

The Sandinistas' development strategy tempered the emphasis on generating exports with an insistence on food security, nutritional well-being, and a more equitable distribution of land and income. The goal of satisfying the basic needs of the poor majority became embedded in efforts to develop the nation's abundant and fertile

---

1. The absence in the Somoza regime of a comprehensive national food policy and the failure to address the problem of hunger and malnutrition are evident in the government economic policy documents of the 1950s through the 1970s. See World Bank, *The Economic Development of Nicaragua* (Baltimore: Johns Hopkins University Press, 1953), pp. 11–14; Consejo Nacional de Economía, *Plan de desarrollo económico y social de Nicaragua, 1965–1969*, pt. 1 (Managua, 1965), pp. 171–174; and Consejo de Planificación Nacional, *Plan nacional de reconstrucción y desarrollo, 1975–1979*, vol. 1 (Managua), pp. 44–47.

land and to stimulate its incipient industry, most of which was devoted to food processing.[2]

The Sandinistas pursued these ambitious goals by entrusting themselves with considerable authority to intervene in the economy. The government nationalized all marketing channels for agricultural exports and proceeded with direct government purchase and sale of basic grains. Rents were controlled. The banking system was nationalized. In addition to assuming these and other levers to guide, cajole, and control the private sector, the state itself became an important producer.

Confiscation of the assets of the Somoza family and their associates gave the new regime approximately 23 percent of the country's most arable land and a vast assortment of highly productive agroindustrial firms, ranging from sugar mills to coffee-processing, milk-pasteurizing, and meat-packing plants.[3] These estates and plants were organized into a hundred-odd state enterprises, collectively named the Area Propriedad del Pueblo, or Area of People's Property (APP). With the creation of these enterprises, the Nicaraguan government became the largest agricultural producer and food processor in the nation. And these state enterprises were given the leading role in implementing Sandinista food policy; they were assigned the crucial tasks of capital accumulation, generation of foreign exchange, and coordination and expansion of the domestic food system.[4]

In the initial years of the revolution, these state agribusiness enterprises partly succeeded in reactivating agricultural output, but after six years it was apparent that they had also engendered a host of unintended results that threatened the economic foundations of the food policy they were meant to support. Among undesired outcomes were the state enterprises' persistent financial losses, their inability to service their heavy and growing debt, the underutilization of their processing plants and equipment, and an unsatisfactory and deteriorating level of labor productivity. The enterprises absorbed a growing amount of ever scarcer national resources and yielded meager returns. By 1985 there was little doubt among the government's economic policymakers that the cumulative cost of this deficient perfor-

---

2. Ministry of Planning, *Programa de reactivación económica en beneficio del pueblo* (Managua: FSLN, 1980).

3. A list of these state-owned agribusiness enterprises, together with a description of their main lines of activity, is provided in the Appendix.

4. Henry Ruiz, *El papel político del APP en la nueva economía sandinista* (Managua: FSLN, 1980).

mance had affected the entire economy by contributing to the significant drop in production and labor productivity, growing plant and equipment inefficiency, black market trading, inflation, chaos in the price structure, and near bankruptcy of the national banking system.[5] According to the government's 1985 economic plan, these deteriorating economic conditions began to endanger not only the food policies these state firms were intended to implement but also the very survival of the revolutionary regime.[6]

Why did these unintended and ultimately disastrous outcomes of state agribusiness enterprises occur, and why did they persist, despite the central government's genuine and repeated efforts to redress the situation?

The argument advanced, elaborated, and supported here is that the ultimate failure of the state enterprises can be explained largely by the government's inability to link effectively the macro dimension of food policy design to its micro, or implementation, level. What was designed and articulated in the capital, Managua, was consistently not feasible at the farms and plants where production actually took place. Sandinista *comandantes*, local administrators, and even field laborers may all have been committed to the broad goals of the revolution. But the specifics, or consequences, of the comandantes' macro policies unintentionally, yet persistently, thwarted the efforts of administrators and laborers.

Macro policies were decreed without a clear sense of their micro consequences. And there was never, in fact, a detailed appreciation of how individual firms, with their gamut of administrators and laborers, responded to fluxes in their managerial environment. The poor performance of individual enterprises was a logical outcome of inappropriate, and often contradictory, policies. Of course, the results of macro policies could reflect none other than the aggregate behavior of individual enterprises.

It is difficult to delineate with any precision why and how macro

---

5. Secretaría de Planificación y Presupuesto, *Plan económico 1985* (Managua: SPP, 1985), pp. 2, 7; E. V. K. Fitzgerald, "Problems in Financing a Revolution: The Case of Nicaragua, 1979–1984," Institute of Social Studies Working Paper (The Hague: Institute of Social Studies, 1985); and Alejandro Argüello, Edwin Croes, and Nanno Kleiterp, "Nicaragua: Acumulación y transformación, inversiones, 1979–1985" (Managua, 1987).

6. SPP, *Plan económico 1985*, pp. 2, 7. The 1985 annual report of the Ministry of Agriculture and Land Reform (MIDINRA), in its assessment of the performance of its state enterprises, also stated that its "negative tendencies must be eliminated; that is, the problem of these [state-owned] enterprises is not a problem of political economy, it is not simply a financial problem, it is a problem of the Revolution" (*Plan de trabajo, balance y perspectivas, 1985* [Managua: MIDINRA, 1985], p. 25).

policies were designed; they appear to have been motivated by a varying assortment of impulses, including vague ideological predispositions, political calculations, attempts to remedy immediate problems, as well as sincere efforts to promote equitable development. It is analytically easier, and perhaps more illuminating, to demonstrate that once formulated, Sandinista macro policies frustrated the noble aims of the revolution. The importance of macro-micro linkage is demonstrated by an in-depth study of four critical macro policies: foreign exchange, food prices, wages, and interest rates. Examining the impact that each of these macro policies had on state enterprises clarifies the inner workings of the food system in revolutionary Nicaragua. The ways in which these policies operated detrimentally at the firm level becomes clear, as does the extent to which they became a serious disincentive to efficient production in state enterprises.

Food policies are difficult to enact even in tranquil times. This task was all the more challenging under the conditions in Nicaragua between 1979 and 1985, the temporal focus of this book. Formulation and implementation of macroeconomic policies were severely constrained by the counterrevolutionary war, which came to absorb 50 percent of the government's budget and a considerable share of the attention of the Sandinista leadership. Nonetheless, the ensuing analysis suggests that most of the difficulties encountered were *independent* of the war. The importance of macro-micro consistency in food policy thus becomes a fundamental lesson of Sandinista-led Nicaragua.

Theoretical Framework

The Sandinista experience with state-owned agribusiness enterprises, frustrating as it was for those who had expectations of greater food security following the revolution, elucidates several important current issues in food policy and agribusiness management in the developing world.

As Peter Timmer explains, the world food crisis of the early 1970s, when international market prices for wheat increased almost fourfold in just eighteen months and prices for rice and soybeans experienced even greater volatility, spawned two sharply divergent schools of thought on economic development.[7] One school, which

---

7. For the analysis in this section, I am heavily indebted to C. Peter Timmer's study, *Private Decisions and Public Policy: The Food Price Dilemma in Developing Countries* (Boston: Harvard Business School, 1984).

came to be recognized as the "basic needs" approach, was concerned with the substantially higher food prices, the apparent scarcity of energy resources, and the decreasing rate of growth in agricultural productivity in the new economic environment. The school turned its attention to the increasingly endangered needs of consumers. It suggested a reorientation of national development strategies toward direct provision of food, housing, education, and health needs as a preferable way to eliminate the worst manifestations of poverty and increase the productivity of the most vulnerable segments of society.[8]

The second school of thought, sometimes called the "production incentives" school, was much more concerned with the broad and persistent tendency of developing countries to undervalue their agricultural sectors and depress food prices. From this perspective, the sudden increases in international prices of basic grains in 1973–74 represented a long overdue readjustment of producers' incentives that would lead, in the long run, to increased supplies.[9]

Both schools of thought were right, but neither was completely so. As Timmer explains:

> The Basic Needs school understood the consequences for food consumption of the new economic environment but failed to provide a policy framework that would lead to sustainable growth in economic productivity in the long run. Its concerns for policy were short run—the timespan in which welfare problems are most pressing. In contrast, the Production Incentives school either missed or ignored the effects higher levels of producer prices, especially for the society's basic foodstuff, would have on food consumption. This school focused attention on policies to promote long-run growth in efficiency and productivity through technological change and trade. But precisely because of its preoccupation with the long run, it missed most of the policy debate over how to help the poor in the short run in the face of scarcer supplies of food and energy.[10]

It was precisely as a conscious effort to bridge these two schools of thought that the focus on food policy emerged in the late 1970s.

---

8. Timmer acknowledges *First Things First: Meeting Basic Human Needs in Developing Countries* by Paul Streeten et al. (Washington, D.C.: World Bank, 1982) as perhaps the most eloquent expression of this perspective and *Food First: Beyond the Myth of Scarcity* by Frances M. Lappé and Joseph Collins (New York: Ballantine, 1979) as a populist version of the same viewpoint.
9. See, for example, Theodore W. Schultz, ed., *Distortions of Agricultural Incentives* (Bloomington: Indiana University Press, 1978).
10. Timmer, *Private Decisions and Public Policy*, pp. 2–3.

The task was to address the "food price dilemma" resulting from the conflict between the short-run and long-run interests of the poor. In Timmer's words:

> In the short run, the poor's primary concern is for low food prices to maintain their access to food. In the long run, their interests are served by incentives that raise economic productivity and generate jobs. Consequently, the dilemma stems from the opposite effects price incentives have on food consumption and production. The debate over food price policy is thus seen as the core analytical problem to address.[11]

The food scare of the early 1970s also had another and potentially broader effect on the research agenda of development economists. By striking a potent blow to the entire international macroeconomic order—an event that, Timmer points out, took place prior to the oil price shocks delivered by OPEC—and by significantly influencing the level of economic activity in many countries, the sometimes overlooked agricultural sector recovered its place in current macroeconomic policy debates.[12] And the renewed interface between agricultural development research, economic development theory, and macroeconomics following the 1974 World Food Conference demonstrated a striking pair of concerns.

On the one hand, as Timmer states, macroeconomists became increasingly aware that they "knew little about an economy beset by sectorial shortages rather than general slack. Microeconomics and commodity modeling suddenly became important to macroeconomics,"[13] as did the agribusiness systems approach.[14] The dynamic effects of agriculture on rural and urban employment, wage structures, investment decisions, and other macro variables raised the question of what micro behavior underlay macro performance, and this question was a central challenge for macroeconomic theory.[15]

On the other hand, recent empirical measurements of the impact

---

11. Ibid., p. 3.
12. Ibid., p. 8. See also Timmer's monograph *The Agricultural Transformation* (Boston: Harvard Business School, 1986). For an example of the growing awareness of the need of agricultural policy to stand on its own and be more initiative than derivative, see John W. Mellor and Richard H. Adams, "The New Political Economy of Food and Agricultural Development," *Food Policy* 2 (November 1986): 289–297.
13. Timmer, *Private Decisions and Public Policy*, p. 8.
14. See, for example, Ray A. Goldberg, *A Concept of a World Food System* (Boston: Harvard Business School, 1984).
15. See George L. Perry, "Reflections on the Current State of Macroeconomic Theory: Reflections on Macroeconomics," *American Economic Review* 74 (May 1984): 401–407.

of macro prices on the food system[16] have provided important evidence supporting the realization that policymakers in the agricultural sector are frequently doomed to fail at efforts to "get agriculture moving" when the macro sector is sending contrary signals.[17] This conclusion rendered unavoidable the task of placing food policy in a macro context.[18]

The interaction between agricultural performance and macroeconomic activity, as well as the role of macro price policies in defining the terms of the food price dilemma, may be better understood in the context of the recent post-Keynesian search to find analytical links between microeconomics and macroeconomics.[19] Although this search has met only partial success, the difficulties of the task are at least becoming clearer. Explains Timmer:

> The links between the micro and macro domains are relatively unexplored because economists have seen few decision makers there with decision rules susceptible to formal analysis. Important actors exist in the area where micro and macro realms converge. Indeed, following Leibenstein's plea that "A Branch of Economics is Missing: Micro-Micro Theory," it is possible to specify a continuum of decision makers [from the micro-micro to macro-macro] in the economic system whose behavior needs to be understood with an economic framework even if that behavior cannot be reduced to simple mathematical optimization models. . . . Behavior at the micro-macro level has been reserved primarily for business schools and, more recently, schools of public policy. Again, the focus has not been on how these actors—the large corporations, unions, cooperatives, public enterprises, and regulatory agencies—relate to the outcome below at micro level and the policies above at the macro level. Rather the analysis has been either descriptive or "strategic"—designing internal mechanisms for coping with what is viewed as a largely exogenous environment at both ends of the spectrum. Consequently, the questions involving the links between these levels, which are mediated by the micro-macro actors, have remained unanswered.[20]

Thus, as Timmer, Walter Falcon, and Scott Pearson conclude, an

---

16. See, for example, C. Peter Timmer, "Macro Prices and Structural Change," *American Agricultural Economics Association* 66 (May 1984): 196–201; and Robert H. Bates, *Markets and States in Tropical Africa* (Berkeley: University of California Press, 1981).
17. Timmer, *Private Decisions and Public Policy*, p. 49.
18. See C. Peter Timmer, Walter P. Falcon, and Scott R. Pearson, *Food Policy Analysis* (Baltimore: Johns Hopkins University Press, 1983), pp. 260–292.
19. Timmer, *Private Decisions and Public Policy*, p. 16.
20. Ibid., pp. 17–20. See also Alan Berg and James Austin, "Nutrition Policies and Programmes," *Food Policy* 9 (November 1984): 309–310.

important contribution of the food policy perspective has been the identification of two dilemmas that are likely to condition any concrete effort to reduce hunger: "a micro-level food price dilemma that recognizes tradeoffs between producer and consumer interest in the short run; and a macro price dilemma that reflects how strongly the macro economy conditions the scope for food policy."[21] The managerial challenges these dilemmas pose to any food policy implementation strategy are far from being fully understood.

Seen from this macro-micro food policy implementation perspective, the performance of Nicaraguan state agribusiness enterprises under Sandinista rule, which has largely been undocumented, is suggestive.[22] First, these enterprises represented a critical implementation instrument of a national food policy that distinguished itself for its widely proclaimed commitment to profit from past experiences. If the focus on food policy as a conceptual approach emerged in the late 1970s as a conscious effort to bridge the "production incentives" and "basic needs" schools of thought, the willingness in post-revolutionary Nicaragua to implement such a food policy was motivated by the Sandinistas' painful awareness of the failure of previous approaches,[23] a failure that appeared all the more difficult to justify in a country with such a rich land base and relatively small population as Nicaragua.

Openly rejecting development models that ignored the fundamentally agrarian and rural nature of the economy, the new Nicaraguan government attempted to meet both the needs of agricultural producers—of which state enterprises represented a significant portion—and those of food consumers. Likewise, the Sandinistas had a clear understanding, particularly from their critique of the Somoza regime, that, as Falcon states, there was "no substitute for treating hunger and other food problems as genuine policy issues and not simply a patchwork of investment projects."[24] To that effect, state

21. Timmer, Falcon, and Pearson, *Food Policy Analysis*, p. 262.
22. Notable exceptions are James Austin, Jonathan Fox, and Walter Krüger, "The Role of the Revolutionary State in the Nicaraguan Food System," *World Development* 13 (January 1985): 15–40; Walter Krüger and James Austin, *Organization and Control of Agricultural State-owned Enterprises: The Case of Nicaragua* (Boston: Harvard Business School, 1983); and Alfred H. Saulniers, *State Trading Organizations in Expansion: A Case Study of ENABAS in Nicaragua* (Managua: Nicaraguan Institute for Economic and Social Research [INIES] / University of Texas at Austin, 1986).
23. See, for example, MIDINRA, *Marco estratégico del desarrollo agropecuario*, vol. 1 (Managua: MIDINRA, 1983), pp. 2–7, and *Marco prospectivo del desarrollo agroindustrial*, vol. 1 (Managua: MIDINRA, 1985), pp. 8–16.
24. Walter P. Falcon, "Recent Food Policy Lessons from Developing Countries," *American Agricultural Economics Association* 66 (May 1984): 182.

enterprises were to assume the leadership of the various agribusiness commodity systems in an all-pervasive, coordinated effort to implement national food policy objectives. The performance of these state enterprises would inevitably have a significant impact on the reconstruction and development of postrevolutionary Nicaragua. The Sandinistas also clearly perceived that by virtue of their double identity as strategic participants in the national food system and direct extensions of the government, these large and critically positioned state agribusiness enterprises could be relied on as the macro-micro link best suited to provide rapid and reliable response to their food policy objectives and to function as ultimate guarantors that the actual behavior at micro food production and distribution levels would be consistent with desired performance.[25]

Unfortunately, through a succession of unanticipated outcomes, state enterprises became partly responsible for the disappointing economic performance of the Sandinista regime. Unraveling what went wrong may provide insights into how, managerially, food policies are best implemented.

Looking back at the work done in the years following the 1974 World Food Conference and the international crisis that prompted it, there is a general consensus that world food problems have not been solved. What has improved, however, is an understanding of hunger and the components of a successful food policy. According to Falcon, "Recognizing these components is itself a lesson . . . [that] is still to be learned in most parts of the world."[26] The Sandinista food policy experiment, and in particular the experience of the state agribusiness enterprises as privileged instruments of the experiment, may facilitate a movement beyond the conceptual phase of policy formulation into an understanding of the administrative challenges of food policy implementation.

Organization of the Book

The temporal focus of this work is on the first five years of the ten-and-a-half-year Sandinista regime, since these were the decisive years for the formulation and implementation of food policies. Be-

---

25. This point is stressed repeatedly in government documents dealing with the APP. It is discussed in more detail in Chapter 2.
26. Falcon, "Recent Food Policy Lessons," p. 181.

ginning in 1985, the economy began to suffer from hyperinflation, and government efforts to control and guide the economy were increasingly frustrated. No new policy initiatives were undertaken during the second half of the decade. The government focused its resources on trying to stabilize an economy that was more and more anarchical. Inflation, for example, reached triple digits in 1985, at 334 percent, and peaked at 33,602 percent in 1988.[27] In 1990, the last year of Sandinista tenure, inflation was 13,491 percent. Concentrating on the period of 1979 through 1985 is most illuminating for those interested in the possibilities and problems of aiding the rural poor majority through ambitious reform. And the failures of Sandinista macro policies during this period help explain the overwhelming economic problems the Sandinistas confronted in the second half of their rule.

Following this opening chapter is a description of the Nicaraguan agricultural sector: its history, evolution, and status on the eve of the revolution; the food policy objectives of the Sandinista leadership; their initial institutional restructuring of the agricultural sector; and the macro performance of the national food system from 1979 to 1985. The ensuing four chapters are an attempt to disaggregate macro performance by examining how specific macro policies affected individual enterprises.

In Chapter 3 I analyze foreign exchange policy by examining its impact on the cotton sector, whose traditional function as the major supplier of foreign exchange, cooking oil, and animal feed to the food system was disrupted by a drastic reduction in production levels. Loyally implementing government policy, but going against the national trend, a large and generally efficient state cotton enterprise increased its cotton production, only to become one of the most unprofitable state enterprises. The cotton pricing system is studied in detail, showing that the enterprise would have been profitable if the overvalued foreign exchange had been adjusted by government-calculated shadow prices.

Chapter 4 explores the consequences of food price policy by studying a once-growing milk and beef sector. Attention centers on the problems confronted by strategically positioned state milk-processing plants and cattle slaughterhouses as they operated at less than half their production capacity. Price policies and the manner in

---

27. Comisión Económica para América Latina (CEPAL), "Notas sobre la economía y el desarrollo," no. 519/520 (December 1991): 39, 42.

which they were implemented are shown to stimulate behavior in state enterprises that undermined their financial viability, contradicted policy objectives, contributed to the deterioration of milk and beef production, depressed consumption and foreign exchange generation, and encouraged the simultaneous emergence of food shortages and a thriving black market. This black market ultimately condemned the same state enterprises to an ugly choice between joining the informal economy, severing whatever food policy commitments that remained, or closing down their operations.

In Chapter 5 I analyze wage rates and their negative impact on labor productivity through an inspection of a state enterprise that produced a basic grain, rice, at a time when growing inflation, food shortages, and the emergence of a highly profitable black market all conspired to make the government's "social wage" unworkable.

Chapter 6 examines interest rate and credit policy through a study of the sugar sector, where production and distribution were predominantly the tasks of state enterprises. I analyze a state sugar refinery, revealing how massive indebtedness and financial losses were the inheritance of an ill-planned expansion of production facilities.

Chapter 7 briefly summarizes how both the central government and individual enterprises tried to cope after 1985 with the most conspicuous manifestation of the country's mounting economic difficulties: hyperinflation. This inflation proved to be so dominating that it became nearly impossible to measure the performance of the state enterprises. Nominal accounting became meaningless. And the ability of the government to formulate macroeconomic policies all but collapsed. For example, the nominal value of the country's currency against the U.S. dollar fluctuated significantly daily. In most enterprises wages came to be renegotiated every other week, with no national, regional, or sectorial consistency. Interest rates were reset every month. There is little doubt that the shortcomings of the revolutionary government's food policies, as well as the general deleterious state of the economy, contributed to the Sandinistas' loss in the national elections held in February 1990. An implicit contention of Chapters 3–6 is that the Sandinista macroeconomic policies of 1979–85 made the overwhelming difficulties of 1985–90 inescapable.

Finally, Chapter 8 offers an integration of the preceding discussions. It reviews the net effects, both economic and political, of permitting a persistent gap between the logic of macro food policies and the microlevel environment where men and women labored. Lessons learned are shared.

12   Hungry Dreams

The Origin of This Book

A Ministry of Agriculture and Land Reform (MIDINRA)/Central American Institute of Business Administration (INCAE) research and training project supported by the Ford Foundation provided access to the intricacies of revolutionary Nicaragua's food policies. Originally conceived, in 1984, as an effort to analyze and improve the planning, budgeting, and control systems of the state agribusiness enterprises, it gradually evolved into an integrated training and consulting project that used the analytical tools of business administration, agribusiness systems analysis, and food policy analysis to address some of the difficulties state managers confronted.

As a result of the active participation of more than two hundred state managers and government administrators and the efforts of a joint MIDINRA/INCAE research team, the project continued through a number of cycles. The intent was to gradually increase the level of complexity of the phenomena studied and treated as understanding of problems progressed. Initial research focused on the financial control problems of state enterprise managers. This work led to the gradual identification of a number of related problems, such as the absence of incentive systems, inconsistent pricing mechanisms, and the practice of unrestricted borrowing, that needed to be addressed if further progress in control were to be achieved. These issues, in turn, had complex policy ramifications that went beyond the perspective of any single enterprise and that were explored and questioned through discussions and seminars before the next research agenda was initiated.

Eventually, as the project's accumulated knowledge reached a critical mass in a number of areas, its main thrust shifted from the design of the state enterprises' internal mechanisms for coping with what state managers came to view as a largely exogenous and threatening environment, to the analysis of the environment itself and how it related to state enterprises and the micro food system they coordinated. If the new task appeared fruitful, it was because both MIDINRA and the research team came to agree that the analysis of entire agribusiness commodity systems, the role of state enterprises in these systems, and the consistency, or lack of it, between macro and sectorial policies and their outcomes at the micro level were at the heart of the state enterprises' problems.

As the project's director from 1984 to 1987, I had the opportunity to interact with most state agribusiness enterprise managers and the numerous government officials assigned to assist or supervise them. I also coordinated the joint efforts of the MIDINRA/INCAE research teams as they attempted to address challenges state enterprises were facing. It was during the course of this activity that, drawing from constructs of agribusiness and food policy theory, I began collecting data and formulating the propositions of this book. Continued contact with senior MIDINRA personnel and other government officials enabled me to keep abreast of the state enterprises until their demise in the aftermath of the 1990 elections in Nicaragua.

I developed the propositions and the unifying argument of this book through an extended iterative process in which I employed the case method as a vehicle for continuous interaction and mutual learning between researchers and practitioners alike while we sifted increasingly wide data bases, gradually identified relevant conceptual frameworks with which to arrange and understand these data, and began to discard earlier unsatisfactory approaches. Propositions were developed a posteriori, as knowledge and understanding of the state enterprises' problems accumulated. Propositions are not tested here in the formal sense. Rather, they are examined in the context of specific cases to establish how much can be said about them, what qualifications need to be made, and what their utility may be for practitioners dealing with the management of food policy implementation. Accordingly, the research methodology is clinical: it uses case research to seek understanding, as opposed to aggregate research, which looks to generalize; experimental research, which strives to replicate; or model building, which aims at predictability.[28]

---

28. For an analysis of these methodological approaches and a conceptual framework regarding research strategies in business administration, see K. Balakrishnan and C. K. Prahalad, *Methods and Strategy for Research in Business Administration* (Boston: Harvard Business School, 1973). See also Andrew R. Towl, "The Use of Cases for Research," in *The Case Method at the Harvard Business School*, ed. Malcolm P. McNair (New York: McGraw-Hill, 1954), pp. 223–229. I am also indebted to John C. Ickis, "Strategy and Structure in Rural Development" (D.B.A. diss., Harvard Business School, 1978), pp. 8–13.

2

Food Policy Formulation: Objectives and Strategy

> The fundamental role of the farm sector is to guarantee adequate food supplies to the population: this is its top priority according to the Revolution. The goal is to achieve self-sufficiency in basic grains by 1990, and in the other dietary components by the year 2000.
> —Nicaraguan Ministry of Agriculture and Land Reform, 1983

> It is not simply a question of increasing the productive capacity of state-owned enterprises in the short and medium term. It is a question of converting them into the strategic sector of the economy.
> —Nicaraguan Ministry of Planning, 1980

In October 1789 Marie Antoinette, queen of France, disturbed by the presence of thousands of protesters outside her residence at Versailles demanding bread and shocked by the news that none was available to appease the hungry crowds, reportedly asked her prime minister, "If you don't have bread, why don't you let them eat cake?" The Queen learned the answer the hard way—she was guillotined. Whether fact or myth, this anecdote exemplifies a basic truth: food problems can erode the legitimacy and stability of even the most established governments. And food policy in the developing world, insofar as it is called upon to address acute food problems, is frequently surrounded by politically explosive issues.

Food policy in revolutionary Nicaragua was not immune to this predicament. Born as a response to the neglected aspirations of the vast majority of the Nicaraguan people under the Somoza regime and rooted in a suffered history of poverty, hunger, and social inequities, it found its immediate identity in an explicit offer of a better life. It is not surprising, therefore, that in the case of Nicaragua, too, as James Austin and Gustavo Esteva observed, "[food] policy formation [occurred] in a bubbling cauldron in which political, economic, demographic, and cultural forces swirl tumultuously."[1] At the same time, no matter how imbued of its nation's turbulent political and social circumstances,[2] Nicaragua's food policy—like that of any nation—presupposes a diagnosis and indicates a treatment that must be followed if cure is to be reached.

Because of the limited scope of the MIDINRA/INCAE research project, the diagnosis and the prescription are taken as given. The primary focus is on the application of the treatment: food policy implementation and the administrative skills that it requires, particularly with respect to state-owned agribusiness enterprises as the government's primary instrument of food policy implementation. Evaluation of the accuracy of the diagnosis and the appropriateness of the prescription, important as it may be, is beyond the scope of this book. This is not to imply that the adopted food policy constituted a definitive answer to the nation's long history of hunger and malnutrition. Even a cursory examination of this policy suggests that some of its premises may have been due for review as the understanding of the nation's complex food system improved. This process, in fact, had already begun, both within and without government institutions. Among the propositions that were being challenged during this writing were that (1) cotton production expanded at the expense of food production,[3] (2) export agriculture was dominated

---

1. James E. Austin and Gustavo Esteva, "The Path of Exploration," in *Food Policy in Mexico: The Search for Self-Sufficiency,* ed. Austin and Esteva (Ithaca: Cornell University Press, 1987), p. 14.

2. The political sensitivity of food policy in revolutionary Nicaragua is illustrated by the variety of available and often conflicting interpretations of its problems. Representative of the spectrum are two articles by E. V. K. Fitzgerald and Marc Falcoff. The former concludes that "the impact of the U.S. aggression . . . has been sufficient to push [Nicaragua's] development back to little more than a subsistence level," while the latter saw the disruptions of the Nicaraguan food system as primarily a consequence of the FSLN's drive for total economic control (E. V. K. Fitzgerald, "An Evaluation of the Economic Cost to Nicaragua of U.S. Aggression, 1980–1984," in *The Political Economy of Revolutionary Nicaragua,* ed. Rose J. Spalding (Boston: Allen and Unwin, 1987), pp. 195–216; and Marc Falcoff, "Nicaraguan Harvest," *Commentary:* 80 [July 1985] 21–28).

3. For example, Francisco Mayorga writes: "It is frequently argued that cotton expanded at the expense of food production: corn, beans, and other food crops were supposedly displaced to marginal lands as cotton took over the better soils. The figures, however, suggest that cotton farming developed through the addition of new lands, under the initiative of medium-sized farmers" (Francisco Mayorga, "The Nicaraguan Economic Experience, 1950–1984: Development and Exhaustion of an Agroindustrial Model" [Ph.D. diss., Yale University, 1986], pp. 42–43).

by a few large landowners,⁴ (3) production of export crops did not significantly contribute to the satisfaction of the basic needs of the population,⁵ (4) the profile of the Nicaraguan food system was determined primarily by the *latifundio-minifundio* relationship (large versus small farms),⁶ (5) the limited industrial base inherited in 1979 was incapable of processing most farm outputs,⁷ and (6) export agriculture introduced a duality in technology between the

4. For example, Peter Marchetti writes: "Recent research shows that small and medium-sized farms controlled almost 40% of agricultural export production: the importance of small farms in the production of coffee enormously increased in the 1970s and came to represent more than half of the output in that commodity. Likewise, small and medium-sized farms came to represent more than 60% of livestock output" (Peter Marchetti, S.J., "Prólogo a la edición en Español," in Solon Barraclough, *Un análisis preliminar del sistema alimentario en Nicaragua* [Geneva: UNRISD, 1984], p. xii).

5. Peter Marchetti points out that one of the "underestimated elements in . . . most studies of agroexporting countries of the periphery is the importance of the agroexporting sector for the satisfaction of [domestic] basic needs. The stereotype of the small underdeveloped country is that of an agroexporting system that prevents the development of production for domestic consumption, and which satisfies primarily the consumption of luxury goods by a dominant class. This type of interpretation needs to be complemented with an analysis of the ways in which agroexport production positions itself at the source of complex production chains which function as pillars of the national economy. The real problem is that the basic consumption needs of the population are tightly interconnected with agroexport. . . . A fall in Nicaragua's agroexports would seriously affect the employment and consumption of the population" (Marchetti, "Prólogo," pp. xiv–xv).

For the argument that in small, trade-dependent Third World economies the external trade sector, rather than peasant food agriculture, is the main source of surplus for accumulation and that foreign exchange, not cheap wage goods, is the key economic constraint, see George Irvin, *Nicaragua: Establishing the State as the Centre of Accumulation* (The Hague: Institute of Social Studies, 1982); E. V. K. Fitzgerald, "The Problem of Balance in the Peripheral Socialist Economy: A Conceptual Note," *World Development* 13 (January 1985): 5–14; and E. V. K. Fitzgerald, "Planned Accumulation and Income Distribution in the Small Peripheral Economy," in *Towards an Alternative for Central America and the Caribbean*, ed. George Irvin and Xabier Gorostiaga (London: Allen and Unwin, 1985).

6. For the argument that the critical analytical relationship that permits an understanding of the Nicaraguan food system is not the large- versus small-farm production interaction but the nature of its rural-urban relationships, see CIERA, *El impacto de la capital en el sistema alimentario nacional* (Managua: MIDINRA, 1984); Marchetti, "Prólogo," pp. xii–xiii; and UNRISD, *Problems of Food Security in Selected Developing Countries* (Geneva: UNRISD, 1986), pp. 190–210.

7. In a major research project, MIDINRA's land reform research center, CIERA, identified several structural impediments to the development of the Nicaraguan food system including "a disarticulated economy, with an industrial base incapable of transforming the majority of agricultural products" (CIERA/PAN/CIDA, *Informe final del proyecto de estrategia alimentaria: Síntesis y conclusiones* [Managua: CIERA, 1984], p. 9). This statement was contradicted by a later MIDINRA study, that analyzed the structure of the agroindustrial sector in much greater depth and noted more accurately that agricultural-processing capacity was not lacking. On the contrary, the study stated that the most relevant characteristic of the agroindustrial sector in the recent past had been the underutilization of its existing plant capacity, primarily because of the insufficient supply of raw materials (MIDINRA, *Marco prospectivo del desarrollo agroindustrial*, vol. 1 [Managua: MIDINRA, 1985], pp. 132–167). See also Chapter 4 of this book.

"modern" export sector and a "backward" production system for the domestic market.⁸

The value of presenting an overview of the nation's food policy objectives and strategy, along with the problems they were meant to solve, resides in the insights they provide into the rationales underlying these objectives and strategies and in their function as a yardstick by which food policy implementation may be evaluated.

The Diagnosis: Hunger in a Land of Plenty

A five-year development plan designed by the World Bank in 1952 at the request of the Nicaraguan government opens with the observation that "few underdeveloped countries have so great a physical potential for growth and development as does Nicaragua."⁹ Indeed, Nicaragua is the largest and least densely populated country of Central America. Its economy has always been, and continues to be, fundamentally agrarian. Its food system permeates all levels of economic activity and generates an estimated 40 percent of its gross domestic product, nearly 60 percent of employment, and over 85 percent of foreign exchange (see Table 1). Richly endowed with land and water resources and exempt from the extreme overpopulation typical of many developing nations, Nicaragua achieved one of the world's highest sustained rates of economic growth during the three decades preceding the 1979 revolution, expanding between the years 1950 and 1977 at an average annual rate of 6.3 percent.¹⁰ Between 1960 and 1979, physical production increased 381 percent in cotton, 335 percent in beef, 210 percent in coffee, and 349 percent in

---

8. Francisco Mayorga states: "It is often argued that the increasing dominance of exports in the agricultural activity introduced an apparent duality in technology: the export sector developed advanced agricultural techniques, making intensive use of imported inputs and machinery; on the other hand, the majority of the farmers producing for the domestic market used rather low levels of technification. The technological dualism, however, was not so straightforward. . . . A more plausible approach to the analysis of the technological advance in agriculture would also have to consider its feasibility in terms of land (flat versus undulating or mountainous, for instance) and the regional weather patterns, which have different implications for different types of crops. Another important consideration is that in the Pacific Lowplains of Nicaragua . . . grain production and even cattle farming were undertaken on the same farms where export crops were cultivated" (Mayorga, "Nicaraguan Economic Experience," pp. 50–52).

9. International Bank for Reconstruction and Development (IBRD), *The Economic Development of Nicaragua* (Baltimore: Johns Hopkins University Press, 1953), p. 3.

10. Central Bank of Nicaragua (BCN), *Indicadores económicos,* vols. 1–4 (Managua: BCN, 1979).

Table 1. Importance of the food system in the national economy

	Value added (1983)		Employment (1982)		Foreign exchange (1982)	
	Million cordobas	%	No. of people	%	US$ million	%
Food system	13,796	40.2	502,087	58.3	353.7	86.8
National economy	34,301	100.0	861,860	100.0	407.7	100.0

Source: CIERA/PAN/CIDA, *Informe final del proyecto de estrategia alimentaria*, vol. 1 (Managua: CIERA, 1985), p. 5.

sugar, while the combined value of exports of these products increased from $40 million to $454 million.[11] Yet this remarkable result benefited primarily a small agrarian oligarchy. Nicaragua in the late 1970s was still a nation of predominantly poor, malnourished, hungry people, whose material conditions and economic prospects were substantially unrelated to the performance of the economy as a whole.[12]

The failure of its otherwise rapidly growing food system to respond adequately to the basic needs of the majority of the population was to prompt a disenchantment among the generation of policymakers who joined the ranks of the revolutionary opposition in the 1960s and 1970s.[13] The scorn with which the new food policy analysts after 1979 frequently referred to this period of overall economic growth—which came to be known as "Somoza's capitalist model of dependent export agriculture,"[14] or more simply as a "re-

---

11. BCN, *Indicadores económicos*, vol. 5, pp. 74–86; and Ministerio de Planificación, *Programa de reactivación económica en beneficio del pueblo, 1980* (Managua: FSLN, 1980), p. 57.

Throughout this book I have used U.S. dollars as the monetary unit of measure unless otherwise indicated. And, unless otherwise indicated, the dollars cited are given for the year in which they were calculated or reported. No indexing or adjustment has been made for inflation of the dollar over the years.

12. James D. Rudolph, ed., *Nicaragua: A Country Study* (Washington, D.C.: Government Printing Office, 1982), p. 37.

13. See, for example, Jaime Wheelock Román, *Imperialismo y dictadura* (Mexico: Siglo XXI, 1975); Jaime Wheelock Román and Luis Carrión, *Apuntes sobre el desarrollo económico y social de Nicaragua* (Managua: FSLN, 1980); and Orlando Núñez Soto, *El somocismo y el modelo capitalista agroexportador* (Managua: Autonomous University [UNAN], 1980), and *El somocismo: Desarrollo y contradicciones del modelo capitalista agroexportador en Nicaragua, 1950–1975* (Havana: Centro de Estudios sobre América, 1980).

14. See, for example, Núñez Soto, *El somocismo y el modelo capitalista agroexportador*.

pressive agro-export model"[15]—reflected the attitude of the postrevolutionary government in rejecting a complex and dynamic food system that had failed in its most fundamental responsibility: to feed its people.[16]

At the time of the downfall of the Somoza regime in mid-1979, Nicaragua had a population of about 2.7 million.[17] Although it had always been predominantly rural, in more recent times it was rapidly urbanizing and growing at an estimated annual rate of 3.3 percent.[18] With its fifty thousand square miles, the country approximates the size of Pennsylvania, but its tropical geography is diversified and commonly divided into four main agricultural regions (see Figure 1). The plains along the Pacific coast, where about 62 percent of the population resides, includes some of the most fertile volcanic soils in Central America and is marked by a succession of rainy (May–October) and dry (November–April) seasons. The Continental Divide, generally referred to as the "interior," consists of the cool, humid mountains and upland plateaus east of the large Lake Managua and Lake Nicaragua. The plains of the Atlantic coast are composed of sparsely populated hot and rainy forests and marshlands. And between the Continental Divide and the Atlantic coast is the "agricultural frontier," a vast and hilly expanse of tropical forests and grasslands in various preliminary stages of agricultural exploitation.[19]

Recent historical research on Nicaragua's food system has identified four basic economic stages.[20] In its simplified form, the first

---

15. Solon Barraclough, *A Preliminary Analysis of the Nicaraguan Food System* (Geneva: UNRISD, 1982), p. 15.
16. See, for example, CIERA/PAN/CIDA, *Informe final del proyecto de estrategia alimentaria*.
17. Nicaraguan Institute of Census and Statistics (INEC), *Anuario estadístico de Nicaragua, 1985* (Managua: INEC, 1986), p. 16.
18. Ibid., p. 16.
19. Barraclough, *Nicaraguan Food System*, pp. 30–31; and FIDA, *Informe de la Misión Especial de Programación a Nicaragua* (Rome: FIDA, 1980).
20. The government's diagnosis of the food system inherited from Somoza in 1979 can be traced in a wide range of government reports or government-sponsored studies. Given chronologically are some of the most significant. Wheelock Román and Carrión, *Desarrollo económico y social*. MIDINRA, *Diagnóstico socio-económico del sector agropecuario* (Managua: MIDINRA, 1980). FIDA, *Informe de la Misión Especial*, pp. 1–64. UNRISD, *Food Systems and Society: The Case of Nicaragua* (Geneva: UNRISD, 1981). Barraclough, *Nicaraguan Food System*, pp. 1–46. MIDINRA, *Marco estratégico del desarrollo agropecuario*, vol. 2 (Managua: MIDINRA, 1983), pp. 1–25. PAN, *Plan quinquenal de alimentación y nutrición, Nicaragua* (Managua: MIDINRA, 1983). Ministry of Internal Trade, *Sistemas de comercialización: Productos básicos de consumo popular* (Managua: MICOIN, 1983). United Nations Children's Fund, *Análisis de la situación económica y social de Nicaragua* (Managua: UNICEF, 1984). CIERA/PAN/CIDA,

Source: Fondo Internacional de Desarrollo Agrícola, *Informe de la Misión Especial de Programación a Nicaragua* (Rome: FIDA, 1980). Reprinted by permission of FIDA.

stage was a colonial economy, whose essential characteristics endured past independence in 1821. Then a succession of coffee, cotton, and cattle booms created a dualistic economy, in which a modern capital-intensive food and fiber export sector (coffee, sugar, beef, bananas, shrimps, irrigated rice, and cotton) coexisted with traditional subsistence farming producing basic wage goods (corn, beans, and nonirrigated rice). The tensions inherent in this export-led

---

Informe final del proyecto de estrategia alimentaria, vol. 1: *El funcionamiento del sistema alimentario* (Managua: CIERA, 1984). CIERA/UNRISD, *Managua es Nicaragua: El impacto de la capital en el sistema alimentario nacional* (Managua: CIERA, 1984). MIDINRA, *Marco estratégico del desarrollo agropecuario regionalizado,* vols. 1, 2 (Managua: MIDINRA, 1985). MIDINRA, *Marco prospectivo del desarrollo agroindustrial,* vol. 1. UNRISD, *Problems of Food Security,* pp. 190–210.

Earlier or prerevolutionary works by Wheelock Román (*Imperialismo y dictadura*) and Núñez Soto (*El somocismo y el modelo capitalista agroexportador* and *El somocismo: Desarrollo y contradicciones del modelo capitalista agroexportador en Nicaragua*) were influential in shaping some of the premises on which later food policy analysis was based.

growth economy created the premises for a structural and political crisis that culminated in the 1979 revolution and the emergence of a new government committed to redress the inherited historical imbalances.

The Colonial Economy, 1520–1870

During the colonial period, an oligarchy of large traditional landowners producing primarily cattle and a few cash crops, such as indigo and cacao, dominated Nicaragua. The country's large native population, estimated at more than 600,000 at the time of the Spaniards' arrival in 1521, was reduced to 30,000 by 1548, mostly by a slave trade that sent a substantial proportion of the population to Peru to be used in the Spanish conquest of the Incas.[21] Landowners reacted in two ways to this drastic depletion of human resources. They turned increasingly to cattle ranching, which was comparatively less labor-intensive than other farming activities and could make use of the abundance of suitable land existing in the Pacific and interior regions. They also tied the remaining labor to the recently created large colonial estates (*haciendas* or *latifundios*) through a system of subsistence food production that, after the indigenous communal lands were broken up, came to be organized in small plots (*minifundios*) near or within the same estates. Export crop production by the haciendas, such as indigo and cacao, complemented this latifundio-minifundio-based economy, introducing some features of a cash economy and exposing the system to the volatility of changing international commodity markets.[22]

Exports of indigo and cacao were variously affected by growing competition from Asia and Ecuador, creating a boom-and-bust pattern. This instability, combined with extensive cultivation practices, absentee landlords, depopulation, and coercive social relations in the production of subsistence food, left a legacy that shaped the evolution of the national food system later in the nineteenth century. By the time it reached independence in 1821, Nicaragua had become the major supplier of beef for Central America, with the cities of

21. See FIDA, *Informe de la Misión Especial*, pp. 1–64; MIDINRA, *Marco estratégico del desarrollo agropecuario*, vols. 1, pp. 1–7, and 2, pp. 1–23; and Jaime M. Biderman, "Class Structure, the State, and Capitalist Development in Nicaraguan Agriculture" (Ph.D. diss., University of California, Berkeley, 1982), pp. 20–46.
22. See FIDA, *Informe de la Misión Especial*, pp. 1–64; MIDINRA, *Marco estratégico del desarrollo agropecuario*, vols. 1, pp. 1–7, and 2, pp. 1–23; and Biderman, "Capitalist Development in Nicaraguan Agriculture," pp. 20–46.

Granada and León as its trade centers. After the cacao and indigo busts, the landed and merchant oligarchy that had previously combined cattle grazing, subsistence food production, and the cultivation of export crops into a dynamic economy, replicated this arrangement with the introduction of coffee on the cattle ranches in the late 1800s. In 1975 Jaime Wheelock, who was later to head MIDINRA, provided the following overall assessment of the colonial period:

> The characteristics of the Nicaraguan economy, shortly before the spread of coffee production, could be summarized in broad terms in the following points: (a) the predominance in the "domestic" economy of a large subsistence sector made of fragmented self-consumption agriculture segregated from the market, and in which the great majority of the economically active population was located; (b) a backward industrial and urban sector with little specialization made of the most poor and marginalized segments of the population; (c) an underdeveloped domestic market and local trade; a lack of integration between agricultural production and industrial craftsmanship; (d) the persistence of the colonial pattern of agroexport—leather, fats and salted beef, indigo, sugar, spices—absorbed and controlled after independence by English, and to a minor extent French and American, trading houses.[23]

The Coffee Era, 1870–1945

Large-scale coffee production began in Nicaragua in the 1890s, several decades after its successful introduction in other Central American countries. This delay was due partly to the resistance of a conservative landed oligarchy that felt comfortable with the ease and prestige of cattle-raising and partly to the continued shortage of rural labor. During the 1870s and 1880s, new laws were introduced that further broke up indigenous communal lands, making more Indian labor available to be exploited in the development of large coffee plantations. With the advent of Zelaya's liberal government of 1893–1909, "liberal" coffee producers found the support they needed to expand coffee production in the face of opposition from "conservative" cattle ranchers, and annual coffee production doubled to eighteen million pounds. The overthrow of Zelaya in 1909 and the subsequent U.S. intervention in Nicaragua strengthened the conservative social forces and led to the development of the predom-

---

23. Wheelock Román, *Imperialismo y dictadura*, pp. 65–67.

inantly large and medium-size coffee plantations of the Carazo uplands and the interior.[24]

Coffee production continued to increase during the first part of the twentieth century and during the 1920s and 1930s came to represent about half of the nation's exports, with cattle, bananas, rubber, timber, cotton, and gold making up the other half. This expansion was facilitated by the passage of vagrancy laws that tied available labor to the land, direct incentives for coffee cultivation, and the continuous incorporation of additional land into large coffee estates that had previously been used by small peasant producers for subsistence farming. Land tenure was increasingly concentrated, the number of landless peasants grew, and basic grains production became further marginalized. The domestic market did not grow, and the majority of the population lived in an ever more impoverished subsistence economy, cultivating basic grains in their small plots and working on the large cattle and coffee estates under conditions of servitude or semiservitude. According to Wheelock, the coffee boom had another important consequence.

> Nicaragua was slowly developing a system of interchange between the "basic grains" economy, cattle raising, and the small industry of the cities and villages. With the irruption of large-scale commercial agriculture the linkages of interdependence between these sectors were broken, partly because the means of subsistence needed by the former basic grains producers that had been displaced into coffee plantations as wage laborers was now coming to a large extent from imports of consumer goods, tools, etc., that the then small domestic production was unable to supply in sufficiently high volume. . . . In this way, the role that the incipient national industry could have then played was played [instead] since the early times by imports, thereby aggravating the dependence of Nicaragua's economy on the dominant international trade centers.[25]

The Cotton Boom, 1945–1965

In the late 1940s and 1950s, the Somoza regime initiated important investments in the infrastructure of the Pacific plains, as well as

---

24. See Wheelock Román, *Imperialismo y dictadura*, pp. 13–48; FIDA, *Informe de la Misión Especial*, pp. 1–64; MIDINRA, *Marco estratégico del desarrollo agropecuario*, vols. 1, pp. 1–7, and 2, pp. 1–23; Biderman, "Capitalist Development in Nicaraguan Agriculture," pp. 20–46; and John C. Ickis and James E. Austin, *Nicaragua: Evolution and Revolution* (Boston: Harvard Business School, 1986), pp. 13–17.

25. Wheelock Román, *Imperialismo y dictadura*, pp. 67–68.

state policies regarding credit and other farm input subsidies favorable to large export-oriented farmers and agricultural processors. The government thereby laid the groundwork for a massive development of cotton production. During the crop years 1951–55, the area planted increased fivefold, to 214,000 acres. By 1978, a record year, the area planted had grown to 541,000 acres. Production between 1950 and 1965 increased at an annual average rate of 25 percent and grew from 22,000 bales in 1951 to 540,000 in 1977–78. On the whole, cotton during this period became the primary source of Nicaragua's foreign exchange, its export value increasing from nearly $2 million in 1950 to $66 million in 1965 and $141 million in 1978.[26]

More than any other event in Nicaragua's agricultural history, the cotton boom vividly confirmed the World Bank's 1952 glowing assessment of the country's economic potential.[27] According to Orlando Núñez, later the director of MIDINRA's Research Center for Economics and Land Reform (CIERA), the expansion of cotton production represented the most significant development of the Nicaraguan agroexport model. It was characterized by (1) a profit and investment pattern that concentrated land ownership in a few hands, (2) the growth of a large and impoverished seasonal labor force, (3) a reduced domestic market, and (4) an incipient industrialization; its overall net effect was to deprive the majority of the labor force of access to land and permanent employment.[28]

Cotton production reinforced the role of the agricultural sector as the country's main pole of investment and its engine of growth, and it gave to the farm all the characteristics of a modern private enterprise. As Wheelock and Carrión explain:

> Cotton production functions within a framework where productive forces are highly developed: intensive use of machinery, application of fertilizers and pesticides, cost and quality control, bank credit, etc. Its annual cycle [as opposed to the longer cycles implicit in coffee and cattle activities]

---

26. BCN, *Indicadores económicos;* National Bank of Nicaragua, *Estudio de la economía del algodón en Nicaragua* (Managua: National Bank of Nicaragua, 1967); Rigo Ordóñez Centeno, *La política crediticia algodonera* (Managua: UNAN, 1976); OEDEC, *El algodón en Nicaragua* (Managua: OEDEC, 1978); Pedro Belli, "An Inquiry Concerning the Growth of Cotton Farming in Nicaragua" (Ph.D. diss., University of California, Berkeley, 1968); DEA, *Datos macroeconómicos de Nicaragua, 1960–1986* (Managua: UNAN, 1987); FIDA, *Informe de la Misión Especial;* and Biderman, "Capitalist Development in Nicaraguan Agriculture."
27. Mayorga, "Nicaraguan Economic Experience," pp. 40–43.
28. FIDA, *Informe de la Misión Especial,* pp. 14–20; and Núñez Soto, *El somocismo y el modelo capitalista agroexportador.*

moreover provides a unique dynamic—in a matter of a few months investments are either recovered with substantial profits or entirely lost—something that gave birth to a modern bourgeoisie that had mastered modern production technologies and was accustomed to the pressures of competition: in short, an entrepreneurial bourgeoisie.[29]

For the large cattle and coffee producers, land ownership had represented the foundation of social prestige; land was an asset worth possessing for its own sake. Now, with the rise of a new competitive system with both high risks and high profits, land ownership lost its medieval significance. During the cotton boom, land came to be perceived as just another commodity, whose profitability depended on its quality and the availability of cheap temporary labor to work it during harvest season.

The cotton boom drastically affected the highly populated and fertile Pacific plains surrounding the urban centers of León, Chinandega, and Masaya. In the 1940s these plains had still been devoted to traditional cattle ranching, with the peasantry using a substantial portion of the land for the production of corn, beans, sorghum, fruits, and other food for domestic consumption. With the advent of cotton, a dramatic process of land dispossession took place, in which the peasantry found itself deprived of their often untitled small plots, fruit trees, and rudimentary houses, to make space for the advancing "white gold." The now landless peasants responded in three ways: some migrated to nearby urban centers, to join the ranks of the underemployed and swell the tide of Nicaraguan urbanization; others moved to marginal lands of the interior, effecting the displacement of basic grains production from the Pacific region to much less developed and less fertile areas; still others offered an organized resistance, as through land invasions, only to be rapidly crushed by the army. In all cases, displaced peasants became increasingly dependent on low seasonal wages paid for cotton and coffee harvesting to alleviate their deteriorating economic conditions.[30]

In short, cotton expansion appears to have occurred at the expense of food production and peasant welfare, continuing on a larger scale the effects of the introduction of coffee. Profits from cotton were invested in developing import operations, mostly related to the

29. Wheelock Román and Carrión, *Desarrollo económico y social.*
30. See Wheelock Román, *Imperialismo y dictadura,* pp. 104–196; FIDA, *Informe de la Misión Especial,* pp. 1–64; MIDINRA, *Marco estratégico del desarrollo agropecuario,* vols. 1, pp. 1–7, and 2, pp. 1–23; Biderman, "Capitalist Development in Nicaraguan Agriculture," pp. 80–110; Núñez Soto, *El somocismo y el modelo capitalista agroexportador.*

acquisition of farm inputs such as fertilizers, pesticides, and agricultural machinery; in creating a variety of activities related to cotton-based ginning and the processing of cooking-oil and animal-feed; in building a distribution network for imported consumer goods attending the needs of a small middle class; and in the gradual creation of a more diversified banking and financial services capability around which several dominant family business groups emerged.[31]

The Beef Export Boom, 1965–1979

Nicaragua began exporting chilled boneless beef to the United States in the late 1950s. Until that time only limited amounts of salted beef and live cattle had been exported, despite the fact that extensive cattle ranching had been a major component of Nicaragua's economy since the early colonial times. A combination of factors appears to have contributed to this new development: the crisis in Argentine beef exports to the United States, an international cattle epidemic that spared Nicaragua and put a premium on its livestock, an increasing world demand for beef, and the availability of new meat-packing and refrigeration techniques that facilitated long-distance transportation.[32]

Once export operations were set in motion, the country's latent potential for cattle breeding and beef production rapidly manifested itself as a new dimension of Nicaragua's export-oriented food system. Imports of high-grade breeding stock, favorable credit terms and technical assistance for cattle improvement programs, financing of slaughterhouses, meat-packing plants, and dairy facilities, and new marketing arrangements for exporting beef, milk powder, and related by-products were soon coordinated by a number of state and private institutions.

This activity eventually generated what food policy analysts after 1979 termed the third phase of Nicaragua's dependent and discriminatory agroexport growth model. During the 1960–79 period the area under pasture more than doubled, reaching 11.6 million acres; the amount of cattle slaughtered more than trebled, from 133,500

---

31. FIDA, *Informe de la Misión Especial*, pp. 1–64; MIDINRA, *Marco estratégico del desarrollo agropecuario*, vols. 1, pp. 1–7, and 2, pp. 1–23; Biderman, "Capitalist Development in Nicaraguan Agriculture," pp. 80–110 and 128–153.

32. FIDA, *Informe de la Misión Especial*, pp. 1–64; MIDINRA, *Marco estratégico del desarrollo agropecuario*, vols. 1, pp. 1–7, and 2, pp. 1–23; Biderman, "Capitalist Development in Nicaraguan Agriculture," pp. 111–127; and MIDINRA, *La ganadería en Nicaragua y sus perspectivas* (Managua: MIDINRA, 1986), pp. 1–9.

head in 1960 to 467,500 head in 1979; and the value of beef exports jumped from less than $3 million to $94 million.[33]

Two consequences of this growth proved later to have important food policy implications. First, the massive expansion of cattle ranching stimulated a further displacement of the population. Small food producers of the cattle-ranching interior, as well as the dispossessed peasants who had recently relocated to this area following the expansion of cotton cultivation in the Pacific plains, were once again dislocated to allow for the expansion of pastures. They migrated to the agricultural frontier to engage in slash-and-burn agriculture in increasingly remote areas. As Jaime Biderman observed: "The typical pattern was for cattle ranchers to allow small food producers to clear a forested area in the agricultural frontier, and to grow food for a year or a few years before expelling them to do the same thing elsewhere, using the newly-cleared land for pasture. Thus, by 1977 it was estimated that the area known as the agricultural frontier contained nearly 20% of the country's rural population, most of them subsisting in very dispersed, exploitative and poverty stricken conditions."[34] It was in this area that the country's worst levels of malnutrition and disease were found.[35]

Second, the development of a beef- and milk-processing industry created an economy much more diversified and articulated than was first recognized by food policymakers in the aftermath of the revolution, and it greatly reinforced the linkages and interdependences between the various commodity systems. Intensive beef and dairy farming in the central and southern Pacific plains became dependent on the availability of animal feed derived from cotton and sugarcane by-products. Cotton oil became a basic ingredient for cooking basic staples such as rice and beans. Hides and animal fat were used for making shoes, soap, and candles.

---

33. BCN, *Indicadores económicos*, vols. 1–5; FIDA, *Informe de la Misión Especial*, pp. 1–64; MIDINRA, *Marco estratégico del desarrollo agropecuario*, vols. 1, pp. 1–7, and 2, pp. 1–23; Biderman, "Capitalist Development in Nicaraguan Agriculture," pp. 111–127; and MIDINRA, *La ganadería en Nicaragua*, pp. 1–9.
34. Biderman, "Capitalist Development in Nicaraguan Agriculture," p. 115.
35. Ibid., pp. 130–141.

The Legacy of Hunger

During the two decades preceding the 1979 revolution, the Nicaraguan economy had been the most dynamic of Central America. Ample evidence indicated that this growth was achieved through an agricultural export-led model that benefited primarily large agricultural entrepreneurs, resulting in a highly skewed rural income distribution (see Table 2) and the impoverishment or the perpetuation of hunger and malnutrition among a large portion of the population (see Table 3). Recalling the experience of the 1960s and 1970s, a U.S. government–commissioned 1982 country study of Nicaragua supported this view when it asserted:

> The economic growth of the past two decades has not at all been matched by corresponding progress in improved living conditions for the bulk of the Nicaraguan population. . . . [Although] per capita GDP . . . from 1960 to 1970 . . . rose 4.2% annually and from 1970 to 1977 [rose] by 2% annually . . . before the FSLN victory [in 1979,] the benefits of Nicaragua's substantial growth had accrued to only a small portion of the population. In 1977 the wealthiest 5% of the population earned about 28% of total income; the poorest 50% earned only about 12% of the total. A skewed income distribution between rural and urban areas also existed. In 1972 the average income in Managua was 3.5 times higher than that in the rural areas. The average income of Managua's families in the lower 50% income bracket was more than 5.5 times higher than that of the corresponding rural group. But within the rural group there was a higher concentration of income; the wealthiest 5% of Nicaragua's farmers earned 42% of income compared to 27% for the corresponding urban group. . . . In 1972 . . . 1.5% of all farms occupied about 40% of the farmland. Furthermore, public programs favored these elites and accentuated the skewed distribution of wealth and income. . . . Rural Nicaraguans consumed an average of only 1623 calories daily [62% of the UN's standard

Table 2. Distribution of rural income, 1971

	Percentage of economically active population	Percentage of income
Medium and large landowners	3.5	63.1
Self-employed workers	45.5	29.4
Other workers	51.0	7.5

Source: FIDA, *Informe de la Misión Especial de Programación a Nicaragua* (Rome: FIDA, 1980), p. 2. Reprinted by permission of FIDA.

requirement] and suffered from serious protein deficiencies—over half the nation's children were malnourished.[36]

The persistence of widespread hunger and malnutrition in the 1970s was a striking feature of Nicaragua's otherwise impressive history of agricultural growth. Yet the problem did not occupy a significant place in the government's policy agenda of that time. Whatever was achieved prior to 1979 by such government institutions as IAN, BNN, INCEI, and INVIERNO in the fields of agrarian reform, rural credit, support prices for basic grains, and integrated rural development projects represented only fragmented and limited efforts incapable of addressing the seriousness of the challenge at hand.[37] A detailed 1976 study by the U.S. Agency for International Development (USAID) underlined this situation when it concluded that, despite the extent of hunger and malnutrition, "up to the present, however, there has been no formal national food and nutrition policy adopted by the government of Nicaragua, nor has any multisectorial policy definition or administrative mechanism been developed to confront effectively the extensive malnutrition existing."[38]

Table 3. Daily per capita consumption of calories, proteins, and fats by income stratum, 1970

	Income stratum			
	Low-income	Medium	High	Very high
% of population	50	30	15	5
Calories	1,767.2	2,703.5	3,255.1	3,931.2
Animal	197.2	337.3	497.8	727.8
Vegetable	1,570.0	2,366.2	2,757.3	3,203.4
Proteins (grams)	46.6	72.5	90.3	111.9
Animal	12.6	22.7	33.8	49.1
Vegetable	34.0	49.8	56.5	62.8
Fats (grams)	31.7	54.2	77.9	114.7
Animal	13.4	22.6	33.3	49.4
Vegetable	18.3	31.6	44.6	65.3

Source: FAO-SIECA, *Perspectivas para el desarrollo y la integración de la agricultura en Centroamérica* (Guatemala: FAO-SIECA, 1974). Reprinted by permission of the Food and Agriculture Organization of the United Nations.

36. Rudolph, *Nicaragua*, p. 37.
37. Biderman, "Capitalist Development in Nicaraguan Agriculture," pp. 128–154.
38. George Pyner and Catherine Strachan, *Nutrition Sector Assessment for Nicaragua* (Managua: USAID, 1976), p. 60.

MIDINRA, in the first comprehensive analysis of the Nicaraguan food system, concluded that eight fundamental factors had prevented the harmonious development of the Nicaraguan food system prior to the revolution. First, the study affirmed, the structure of production was centered around farm exports. These, in turn, were controlled by a small social group and were developed at the expense of the production of basic grains for domestic consumption, a task to which the great majority of small farmers were devoted. Second, the economy and production technologies adopted were highly dependent on imports of farm inputs, raw materials, and capital and consumer goods, creating an agribusiness economy highly vulnerable to the volatility of international commodity markets and deteriorating terms of trade. Third, consumption was highly biased in favor of the urban sector and the wealthier segments of society. Profit margins of food producers were minimal, and distributors captured most of the economic benefits accruing from the food system. Fourth, the economy's infrastructure was concentrated in a few areas, leaving entire regions and their populations completely marginalized. Fifth, poor use was made of the abundant natural resources, particularly water and land, and production of certain basic grains had been misplaced to areas where the climate was less appropriate to their cultivation. Sixth, the majority of the population was concentrated in the urban centers, thus preventing the farm sector from making efficient use of the country's already scarce human resources. Seventh, the structure of the economy was further disarticulated because of the presence of a limited industrial base, which was incapable of processing most farm outputs, and an overexpanded tertiary and informal sector providing unstable and partial employment. Eighth, the country had a poor institutional capacity for economic planning, and there was a lack of basic analysis of the nation's resources and development potential.[39]

This diagnosis, influenced by Wheelock's early writings and developed and refined after 1979 through a succession of government studies, is presented here in only its essential elements. The analysis identified the main forces that had shaped the country's changing food system over its long history, provided an explanation for the persistence of widespread hunger and malnutrition, and served as a basis upon which Nicaragua's postrevolutionary government proceeded to formulate the nation's first comprehensive food policy.

---

39. CIERA/PAN/CIDA, *Informe final del proyecto de estrategia alimentaria*, pp. 7–10.

Food Security: A Unifying Objective

According to MIDINRA, the two fundamental goals of the new government were the "strengthening of national sovereignty," defined as the political autonomy needed to implement a national program of great social and economic transformations, and the "transition to a New Economy," a society where peasants and other poor workers, who comprised the vast majority of the population, would have the leading role in generating the nation's wealth and in benefiting from it.[40] Given the predominance that the agricultural sector and its related food-processing industry had in the national economy, MIDINRA's task of translating the two national goals into sectorial objectives in effect defined the agenda for national development.

MIDINRA identified two basic objectives. First was food security, understood as the combination of food access in adequate quantity and quality for all the population, and national self-sufficiency in food supply.[41] Second was the generation of foreign exchange, a critical ingredient for future economic development. This goal was to be achieved through further diversification of the agricultural production and export markets, an increase in the volume of existing export crops, and expansion of a food and fiber industry oriented toward both domestic and export markets that would augment the value added of domestic farm products.[42]

These two objectives were not conceived as separate entities of a sectorial program but were viewed instead as interdependent and complementary elements of the larger goal of achieving a more equitable society, in which food security represented the first priority and its unifying operating principle. In the words of Wheelock, the head of MIDINRA since its inception in 1980, "The food problem is acute, and whatever path the Revolution takes, from the point of view of its own development, it has to resolve first the problem of the basic consumption needs of the population; that is the first priority which we have given to the Revolution."[43] Thus, the generation of increased foreign exchange was considered essential for im-

---

40. MIDINRA, *Marco estratégico del desarrollo agropecuario*, vol. 1, p. 7.
41. Ibid., pp. 17–23.
42. Ibid., pp. 24–30.
43. Jaime Wheelock Román, *El gran desafío* (Managua: Editorial Nueva Nicaragua, 1983), as quoted by James E. Austin and Jonathan Fox, "Food Policy," in *Nicaragua: The First Five Years*, ed. Thomas W. Walker (New York: Praeger, 1985), p. 399.

porting the capital and intermediate goods needed for setting the base of future economic development, but because that development included increased production of food and other wage goods as one of its essential components, exports no longer were viewed as a conflict to food policy. Likewise, food security implied not only the improvement of consumers' income but also an increase in food production, which in turn depended on the availability of foreign exchange—and agricultural exports—for its achievement. Physical interdependencies between domestic and export production, such as those existing between cattle, basic grains, and cotton farming, only served to reinforce the complementarity of these objectives.[44]

Defining Food Consumption Needs and Food Production Targets

Food security had as its central goal the achievement of self-sufficiency in basic grains by 1990 and in the other dietary components by the year 2000.[45] On the consumption side, food security entailed access to a recommended minimum diet, calculated by age and sex, and amounting to a daily average per capita food intake of 2,242 calories and 70 grams of protein, plus vitamins and minerals. Estimates of total food demand for the years 1985, 1990, and 2000, along with the corresponding increases in domestic food supplies needed to satisfy it, are shown in Table 4. This demand was calculated on the basis of population projections and dietary requirements made by MIDINRA with the assistance of the Ministries of Health, Labor, and Planning.[46] On the supply side, increases in basic grains production needed to achieve self-sufficiency were relatively modest, with annual growth rates varying between 1 percent and 6.5 percent during the period 1981–90 and decreasing thereafter. Much greater increases, however, were needed for meats, vegetables, eggs, and milk. Accordingly, MIDINRA's food production targets were set so as to enable the country to reach self-sufficiency in basic grains by 1990 and in the other dietary components by the year 2000, as reported in Table 5.

---

44. MIDINRA, *Marco estratégico del desarrollo agropecuario*, vol. 1; MIDINRA, *Marco estratégico del desarrollo agropecuario regionalizado*, vol. 2; and MIDINRA, *Marco prospectivo del desarrollo agroindustrial*, vol. 2, pp. 19–141.
45. MIDINRA, *Marco estratégico del desarrollo agropecuario*, vol. 1, pp. 19–23; and MIDINRA, *Marco estratégico del desarrollo agropecuario: Resumen ejecutivo* (Managua: MIDINRA, 1983), p. 7.
46. MIDINRA, *Marco estratégico del desarrollo agropecuario*, vol. 3, pp. 1–27.

Table 4. Projected food demand

Product	Net supply[a] (000 tons) 1981	Food demand (000 tons) 1985	1990	2000	Production[b] (average annual growth rates) 1981–85	1981–90	1981–2000
Livestock							
Poultry	7.3	23.5	27.6	37.6	33.9%	15.9%	9.0%
Pork	6.0	16.0	18.8	25.6	27.8%	13.5%	7.9%
Beef	23.8	30.9	36.4	49.6	6.7%	4.8%	3.9%
Eggs	9.3	32.6	38.3	52.2	36.8%	17.0%	9.5%
Milk	150.0	420.4	494.0	674.4	29.4%	14.2%	8.2%
Basic grains							
Rice	87.4	81.1	95.3	130.1	(1.8%)	1.0%	2.1%
Beans	53.3	80.2	94.2	128.6	10.8%	6.5%	4.7%
Corn	170.2	241.8	288.5	403.1	9.1%	6.0%	4.6%
Sorghum	25.1	25.1	25.1	25.1	0.0%	0.0%	0.0%
Other							
Vegetables	48.2	130.6	153.4	209.4	28.3%	13.7%	8.0%
Roots	37.5	62.3	73.2	99.9	13.5%	7.7%	5.3%
Fruits	51.0	116.5	136.9	186.9	22.9%	11.6%	7.1%
Sugar	126.0	165.2	194.1	265.0	7.0%	4.9%	4.0%
Fats	16.7	34.4	40.4	55.2	19.8%	10.3%	6.5%

Source: MIDINRA, *Marco estratégico del desarrollo agropecuario*, vol. 1 (Managua: MIDINRA, 1983), p. 22.
[a]Corresponds to actual production less storage losses and animal consumption. In the case of beef and sugar, exports are not included.
[b]Corresponds to production growth rates needed to satisfy 100% of food demand.

Table 5. Production targets for reaching food self-sufficiency in the year 2000 (000 tons)

Product	1985	1990	2000
Livestock			
Poultry	10.3	15.9	37.6
Pork	8.1	11.9	25.6
Beef	27.8	33.7	49.6
Eggs	13.4	21.1	52.2
Milk	205.8	305.7	674.5
Basic grains			
Rice	83.9	98.5	137.2
Beans	75.1	101.5	139.8
Corn	229.7	324.9	458.5
Sorghum	26.7	26.7	26.7
Other			
Vegetables	87.6	120.8	261.8
Roots	56.2	70.1	117.5
Fruits	78.8	111.1	219.9
Sugar	152.6	185.4	279.6
Vegetable oil	22.7	30.7	58.6
Animal feed (sorghum)	94.8	163.3	439.1

Source: MIDINRA, *Marco estratégico del desarrollo agropecuario*, vol. 1 (Managua: MIDINRA, 1983), p. 23.

Defining Import Needs and Agricultural Export Targets

The availability of foreign exchange had always been an important condition for fueling the nation's economic growth. If the 1979 revolution brought any change in this respect, it was in the direction of making foreign exchange an even more critical resource.[47] Added to the need for traditional imports, such as capital goods and farm inputs used to produce agricultural exports, were new import requirements necessary to sustain a policy of increased food production, as well as a demand for pharmaceutical products, public transportation vehicles, and other equipment needed to carry out improvements in social services. To this increased demand for foreign exchange also had to be added some contingent necessities that had surfaced in the aftermath of the revolution. Food and other consumer goods had to be imported temporarily to compensate for the drop in production caused by the armed struggle in 1979 and to

[47]. See E. V. K. Fitzgerald, *Problems in Financing a Revolution: The Case of Nicaragua, 1979–1984* (The Hague: Institute of Social Studies, 1985), p. 9; E. V. K. Fitzgerald, "Economic Problems in the Analysis of Transition in Nicaragua," in *The Transition Strategy of Nicaragua*, ed. Ruerd Ruben (Amsterdam: Free University, 1982), pp. 57–66; Fitzgerald, "The Problem of Balance," pp. 5–14; Fitzgerald, "Planned Accumulation and Income Distribution"; and Irvin, *Nicaragua: Establishing the State*.

meet the increasing consumption expectations that new government policies had raised among the majority of the population. New machinery and equipment had to be purchased to replace what had been damaged during the war. The servicing of the foreign debt left by the Somoza administration, amounting to approximately $200 million annually, had to be paid.[48]

In this respect, the continuation and further increase of the country's traditional agricultural exports represented the only realistic way out of the impasse. Nicaragua had a proven ability to export farm products, and their importance in the trade balance was overwhelming. In 1979 and 1981, about 80 percent of total exports came from agriculture, with four commodities—coffee, cotton, beef, and sugar—representing almost 90 percent of that amount.[49] As a result, the objective of generating foreign exchange called for more exports of agricultural products, with an eye on product and market diversification as a way of reducing trade risks.[50]

Projected foreign exchange needs were estimated at $1.2 billion in 1990 and $1.9 billion in the year 2000 on the basis of an evaluation of the country's minimum import requirements. After considering a number of export alternatives under different price, market, and product mix assumptions, the agricultural and agroindustrial sectors were assigned to cover about 65 percent of those foreign exchange needs, and the corresponding estimated physical production targets necessary to meet those needs were set for fourteen selected agricultural export products (see Table 6).

Strategy: State Enterprises as Food Policy Implementers

According to government policy analysts, what would distinguish Nicaragua's strategy for attaining food security from similar efforts elsewhere was "the fact that [in Nicaragua] the problem is confronted by attacking its causes not in a partial way, but in all its dimensions."[51] Austin and Fox noted that the new government began to address the inherited structural economic problems in the consumption and production sides simultaneously.[52]

---

48. MIDINRA, *Marco estratégico del desarrollo agropecuario*, vol. 4, pp. 1–61.
49. Ibid., pp. 4–5.
50. Ibid., pp. 8–53.
51. CIERA, *Informe del primer seminario sobre estrategia alimentaria* (Managua: MIDINRA, 1983).
52. Austin and Fox, "Food Policy," p. 400. Throughout this section I am heavily

Table 6. Export targets for the agricultural sector (000 cwt)

Product	1980–82	1989–90	1999–2000
Cotton	1,377.6	2,150.0	3,150.0
Coffee	1,147.0	2,000.0	3,500.0
Sugar	2,078.0	4,295.0	5,594.0
Tobacco	21.0	380.4	380.4
Havana	4.9	10.0	10.0
American	3.0	70.4	70.4
Burley	13.2	300.0	300.0
Bananas	2,074.8	2,856.0	2,856.0
Sesame	138.0	750.0	800.0
Cacao	0.042	117.9	117.9
Citrus	2.7	610.0	2,500.0
Vegetables			
Canned	—	270.6	270.6
Fresh	—	600.0	1,000.0
Rubber	—	142.0	712.0
Lumber	5.2[a]	95.2[a]	190.4[a]
Other	80.0	207.3	414.7
Beef	201.8	820.2	1,578.8
Vegetable oil	—	233.4	583.6

Source: MIDINRA, *Marco estratégico del desarrollo agropecuario*, vol. 1 (Managua: MIDINRA, 1983), p. 32.
[a]Thousand cubic meters.

---

indebted to the just-mentioned article by Austin and Fox, as well as to that by James E. Austin, Jonathan Fox, and Walter Krüger, "The Role of the Revolutionary State in the Nicaraguan Food System," in *World Development* 13 (January 1985): 15–40.

Given next are the primary sources for this section. FIDA, *Informe de la Misión Especial*. UNRISD, *Food Systems and Society*. PAN, *Programa prioritario de la Revolución: Programa alimentario nacional* (Managua: MIDINRA, 1981). MIDINRA, *Marco prospectivo del desarrollo agroindustrial*, 3 vols. Barraclough, *Nicaraguan Food System*. MIDINRA, *Marco estratégico del desarrollo agropecuario*, 6 vols. PAN, *Plan quinquenal de alimentación y nutrición*. MICOIN, *Sistemas de comercialización: Productos básicos de consumo popular*, 6 vols. CIERA, *La situación del abastecimiento*. (Managua: MIDINRA, 1983). CIERA, *Distribución y consumo popular de alimentos en Managua* (Managua: MIDINRA, 1983). CIERA, *Informe del primer seminario sobre estrategia alimentaria*. MIDINRA, *Informe de Nicaragua a la FAO, 1983* (Managua: MIDINRA, 1983). PAN, *Plan contingente para la producción de granos básicos (Año I)*, 2 vols. (Managua: MIDINRA, 1983). PAN, *Propuesta programa alimentario avícola–pecuario* (Managua: MIDINRA, 1983). MIDINRA, *Producción, distribución, y consumo de los principales productos de consumo básico* (Managua: MIDINRA, 1983). PAN, *Problemática del abastecimiento de productos hortícolas en Nicaragua* (Managua: MIDINRA, 1984). PAN, *Estrategia alimentaria en Nicaragua* (Managua: MIDINRA, 1984). CIERA, *Informe borrador del proyecto de estrategia alimentaria* (Managua: MIDINRA, 1984). Directorate General of Agriculture (DGA), *Informe sintético evaluativo de la cosecha en el plan contingente de granos básicos y perspectiva de desarrollo* (Managua: MIDINRA, 1984). PAN, *Informe global del plan contingente para la producción de granos básicos* (Managua: MIDINRA, 1984). PAN, *Perfil del proyecto "Granjas Integrales para la Seguridad Alimentaria Rural"* (Managua: MIDINRA, 1984). PAN, *Plan quinquenal de alimentación y nutrición, Plan global, 1984–85/1989–90* (Managua: MIDINRA, 1984). PAN, *Plan quinquenal de alimentación y nutrición, Planes de acción 1984–1985* (Managua: MIDINRA, 1984). PAN, *Propuesta*

Consumption: Ensuring Access to Food

On the consumption side, the strategy consisted of finding mechanisms that would ensure economic and physical access to adequate food supplies for the population.[53] Economic access meant primarily that the consumers' purchasing power had to be improved. This goal was to be achieved through a combination of price controls on basic staples, employment generation programs that would make available relatively inexpensive food while they increased the income of the poorer segments of society, and consumption subsidies that would resolve the basic food policy dilemma of trying to maintain both low prices for consumers and price incentives to producers.[54] Physical access was just as important, for without it economic gains could not be translated into actual food consumption increases. Strategy at this level focused on food distribution at its critical stages: imports, domestic procurement, storage, wholesaling, and retailing.[55]

Food imports were to satisfy a portion of the rapidly increasing food demand until domestic production recovered from the 1979 drop in output and from other disruptions brought by the war. The other measures were meant to ensure that sufficient quantities of a growing domestic food production would be channeled to low-

---

metodológica de balances alimentarios de granos básicos y hortalizas (Managua: MIDINRA, 1985). PAN, *Estrategia para la orientación del sistema alimentario nicaragüense* (Managua: MIDINRA, 1985). PAN, *Problemática de la producción del frijol y propuestas de medidas estratégicas e inmediatas para su confrontación* (Managua: MIDINRA, 1985). PAN, *Plan de emergencia* (Managua: MIDINRA, 1985). PAN, *Líneas generales. Plan emergente. Programa de movilización popular para el autoabastecimiento alimentario* (Managua: MIDINRA, 1985). PAN, *Evaluación del programa de movilización popular para el autoabastecimiento alimentario 1984 y líneas generales de trabajo 1985/86* (Managua: MIDINRA, 1985). PAN, *Historial de la formulación y gestión del proyecto "Apoyo a la Seguridad Alimentaria a través del P.A.N."* (Managua: MIDINRA, 1985). PAN, *Plan nacional operativo de producción de alimentos básicos, ciclo agrícola 1983/84* (Managua: MIDINRA, 1983). PAN, *Plan nacional operativo de granos básicos 1983/1984* (Managua: MIDINRA, 1983). CIERA, *Distribución y consumo de alimentos populares en Managua* (Managua: MIDINRA, 1983). UNICEF, *Análisis de la situación económica y social de Nicaragua*. CIERA/PAN/CIDA, *Informe final del proyecto de estrategia alimentaria*, 4 vols. CIERA/UNRISD, *Managua es Nicaragua: El impacto de la capital en el sistema alimentario nacional*. MIDINRA, *Marco estratégico del desarrollo agropecuario regionalizado*, 4 vols. UNRISD, *Problems of Food Security*, pp. 190–210.

53. Austin and Fox, "Food Policy," pp. 400–405.
54. Austin and Fox, "Food Policy," pp. 400–401; and CIERA/PAN/CIDA, *Informe final del proyecto de estrategia alimentaria*, vol. 3: *Directorio de políticas alimentarias*, pp. 59–73.
55. Austin and Fox, "Food Policy," pp. 401–405; and CIERA/PAN/CIDA, *Directorio de políticas alimentarias*, pp. 74–87.

38   Hungry Dreams

income urban and rural consumers at regulated prices. These measures included the establishment of a centralized food procurement capability, the construction of more food storage capacity, the creation of new "secure" wholesale and retail distribution channels, the formation of a network of neighborhood food inspectors, and the mobilization of unions and other mass organizations.[56]

Production: Increasing Domestic Food Supply

The effort to reach complete food self-sufficiency prompted measures aimed at raising the domestic output of a wide range of food products (see Table 5). Some of these products, such as beef, sugar, rice, and powdered milk, had been regularly exported in the past. Others, particularly poultry, pork, eggs, pasteurized milk, and cooking oil, as well as beef, depended for their continued production on the availability of animal feed and other by-products, such as cotton cake meal and cottonseed, that were derived from export crops. And all these food products needed varied amounts of foreign exchange to fuel their planned expansion. The larger goal of achieving food security took into account the interdependence of domestic and export agriculture and the fact that any strategy aimed ultimately at increasing domestic food supply must be directed to both domestic and export production.[57] The strategy proposed involved a package of eleven food programs, special policies in the areas of pricing, credit, investment, and technology, and changes in the production structure.

Food Programs. Among the most important food programs were the National Food Program (PAN), whose strategy was directed at "achieving food security for the Nicaraguan people through self-sufficiency in basic grains and the creation of a distribution and commercialization system based on the interests and participation of the masses";[58] the Plan for Livestock Development, which aimed at increased milk and beef production and planned the development of the two largest dairy farms in Central America, four cattle-breeding

---

56. Austin and Fox, "Food Policy," p. 402; CIERA/PAN/CIDA, *Directorio de políticas alimentarias,* pp. 74–87; and MIDINRA, *Marco estratégico del desarrollo agropecuario,* vol. 1, pp. 10–65.
57. Austin and Fox, "Food Policy," pp. 404–408; CIERA/PAN/CIDA, *Directorio de políticas alimentarias;* MIDINRA, *Marco estratégico del desarrollo agropecuario,* vol. 1, pp. 10–65.
58. PAN, *Programa prioritario de la Revolución,* p. 2.

ranches, and three animal feed plants; the Plan for Sugar Development, which included the construction of the region's largest sugar mill, as well as the expansion and modernization of other existing mills; the Plan for Vegetable Development, which included creating an integrated 3,000-acre complex for vegetable growing and processing and significantly expanding an existing operation for tomato and tropical fruit growing and canning; and the Plan for Vegetable Oils Development, which included the planting of a total of 15,000 acres of African palms and 2,600 acres of coconut trees in five locations on the Atlantic coast and the agricultural frontier, for vegetable oil production.[59]

Price Incentives. Price incentives to producers, according to a CIERA/PAN/CIDA report, were an essential element of an overall "pricing policy through which the state, after the revolution, aimed on the one hand to stimulate production as an effort to maintain and expand the supply of food and foreign exchange, and on the other hand to maintain and expand consumers' purchasing power."[60] These incentives included (1) "cost-plus" guaranteed domestic prices for export crops, partially convertible into dollars, that were meant to relieve farmers from the risks of international price fluctuations and assure them of an acceptable profit margin; (2) subsidies on imported farm inputs through a mechanism of differentiated foreign exchange rates; and (3) price supports for basic grains.[61]

Credit Policy. Credit policy was meant to assure the continuity of previous financing policies toward large private landowners and at the same time provide massive support to both the newly created state farms and cooperatives and those producers who had been traditionally neglected by the previous regime, particularly the small and medium-size farmers.[62]

Investment Policy. Investment policy had as its primary function the establishment of an expanded productive basis for future increases in the output of food and export. Policy strategy was guided by three principles. The first stated that future increases in foreign exchange, employment, and the generation of income had to come through the industrialization of agriculture. The second ex-

59. CIERA/PAN/CIDA, *Directorio de políticas alimentarias*, pp. 1–21.
60. Ibid., p. 59.
61. Ibid., pp. 59–73.
62. Ibid., pp. 39–50.

plained that priority within this industrialization process had to be given to those projects that expanded the supply of basic consumer goods, particularly processed foods, that were likely to bring the country closer to its stated objective of food self-sufficiency. The third principle stated that investment policy had to develop new sources of financial and economic aid with countries that offered favorable terms and did not impose constraints on national sovereignty.[63]

Technology Policy. Policy regarding technology reflected the situation inherited in 1979, in which farms were broadly divided according to three different means of production. Modern capital-intensive production was represented by the new state-owned enterprises (mostly consisting of Somoza's former landholdings) and those large private farms that had not been expropriated. Extensive agriculture was practiced mostly by cattle ranches. Traditional agriculture was represented by the bulk of small and medium-size farms and cooperatives that used little mechanization or modern farm inputs in their activities. Policy was oriented, first, toward consolidating and spreading existing modern technology in the less advanced sectors and reorganizing the institutional capability for farm extension programs. Second, it promoted large-scale, capital-intensive, irrigated projects for sugar, dairy, vegetables, and basic grains, which were to spearhead food production growth.[64]

Changes in Production Structure. One of the first measures taken in 1979 by the new revolutionary government was the expropriation of Somoza's assets, along with those of his family, close allies, and high-ranking military personnel.[65] With that decision the state became, overnight, the largest landowner and food producer in the country, gaining direct control over an estimated 23 percent of the nation's cultivated land—eventually subdivided into approximately one hundred state farms—and the owner of a vast assortment of highly productive agroindustrial firms ranging from sugar mills to coffee-processing, milk-pasteurizing, and meat-packing plants, which together with state farms came to be designated collectively as the Area of People's Property, or APP (for a listing of APP's firms and their activities, see the Appendix).

63. Ibid., pp. 22–31.
64. Ibid., pp. 51–58.
65. Decree nos. 3, 38, 580, and 696 of August 22 and September 3, 1979, December 9, 1980, and April 7, 1981.

Although little disagreement was expressed over the expropriation of Somoza's assets, which were generally regarded as consisting mainly of "stolen" property,[66] the decision to organize them into state-owned enterprises was a distinct strategic choice. Although not immune from external criticism,[67] this move seemed within government circles to be entirely consistent with stated food policy. Government officials pointed out that past economic growth had been generated mainly by private commercial farmers and agribusiness entrepreneurs and that this growth had proved unable to alleviate the hunger and malnutrition problems of the population. Accordingly, they argued that more of the same growth, in itself, would not lead to food security. Instead, state leadership at each stage of the food system, including production, was thought to be needed to reach stated food policy objectives.[68] Large and dynamic agribusiness entrepreneurs were viewed with suspicion by a government that attributed the persistence of poverty, hunger, malnutrition, and general economic and social inequities to a perceived negligence or outright carelessness on the part of these businesspeople. The fact that Somoza and his allies had left behind substantial agribusiness assets that could be transformed by a stroke of a pen into state-owned enterprises appeared to be a fortuitous circumstance, under which state leadership could be translated into an immediate reality at little apparent social and political cost.[69]

Actually, the expropriation of Somoza's holdings and the creation

---

66. See, for example, "The Historic Program of the FSLN, 1969," in *Conflict in Nicaragua: A Multidimensional Perspective,* ed. Jiri Valenta and Esperanza Durán (Boston: Allen and Unwin, 1987), p. 322, which advocated the expropriation of all the assets of the Somoza family because they were "accumulated through the misappropriation and plunder of the nation's wealth"; see also Austin, Fox, and Krüger, "Role of the Revolutionary State," p. 16.

67. See, for example, Alejandro Bolaños Geyer's critical assessment, "La economía mixta en Nicaragua," in *La economía mixta en Nicaragua,* ed. CINASE (Managua: CINASE, 1986), as well as the many reports published on this topic by the private sector's organization COSEP.

68. See, for example, Henry Ruiz, *El papel político del APP en la nueva economía sandinista* (Managua: FSLN, 1980).

69. The apparently fortuitous nature of the circumstances surrounding the creation of the Nicaraguan state-owned agribusiness enterprises seems to be confirmed by Jaime Wheelock's declaration that "we were not choosing a model [of state ownership], the model was chosen for us by the realities" (as quoted by Joseph Collins in *Nicaragua: What Difference Could a Revolution Make? Food and Farming in the New Nicaragua,* 3d ed. [New York: Grove Press, 1986], p. 60). Among the government's reasons for choosing state ownership were, according to Collins, (1) the Sandinistas' fear that if the confiscated properties were parceled up, productivity would drop; (2) the belief that parceling out the properties would inevitably be unjust (how would it be decided who should get a parcel?); (3) the belief that it would be easier to create more jobs on large state farms than on small

of the first state enterprises triggered a more complex process of change in the structure of agricultural production. The framework of this process, at least in its general terms, had actually been conceived years before.[70] The stated objective was to make the economy more responsive to what was termed the "logic of the majority," as opposed to the logic of individual profit maximization. This objective entailed state ownership of critical food system components, land expropriations and redistribution, the cooperativization of small-farm producers, changes in land use, and other measures intended to ensure that the nation's wealth would "percolate up" rather than "trickle down," as had been the case in the past.[71]

Accordingly, agrarian policy after 1979 envisioned five legitimate participants in the farm sector but assigned them different dynamics. Large private landowners, who represented properties of more than

---

parcels worked by family labor; (4) the belief that state ownership was the fastest way to revive agricultural production; and (5) the fear that if peasants were granted direct control of the farms, almost all of which were set up to produce for export, they would plant food crops instead, preventing the restoration of the nation's capacity to earn foreign exchange (pp. 59–62). Others have questioned these reasons and have emphasized the FSLN's quest for absolute power through the expansion of state economic control over the means of production (see, for example, David Nolan, *The Ideology of the Sandinistas and the Nicaraguan Revolution* [Miami, Fla: University of Miami, 1985]; Falcoff, "Nicaraguan Harvest," pp. 21–28; Arturo J. Cruz, Sr., "Leninism in Nicaragua," in *Conflict in Nicaragua,* ed. Valenta and Durán, pp. 41–52; and Bolaños Geyer, "La economía mixta en Nicaragua." Whatever the true intentions, the creation of state-owned enterprises was not fortuitous if that means it was an unpremeditated or tactical decision born on the spur of the moment (see n. 70).

70. See, for example, (1) the economic measures advocated in the 1969 Historic Program of the FSLN, which included the expropriation of all the landed estates, factories, companies, buildings, means of transportation, and other wealth owned by the Somoza family and by the politicians and military officers associated with the Somoza regime; the nationalization of the banking system; state control over foreign trade; and central planning of the national economy, "putting an end to the anarchy characteristic of the capitalist system of production" ("The Historic Program of the FSLN, 1969," in *Conflict in Nicaragua,* ed. Valenta and Durán, pp. 321–329); (2) the structural economic transformations sought by Carlos Fonseca, founder of the FSLN, and described in his writings during the 1960s and early 1970s (Fonseca, *Obras: Bajo la bandera del Sandinismo,* vol. 1 [Managua: Editorial Nueva Nicaragua, 1982]); and (3) the emphasis placed by the Ministry of Planning, since its inception in 1979, on state control of the means of production (see, for example, Henry Ruiz, *El papel político del APP;* and Eduardo Baumeister and Oscar Neira Cuadra, "The Making of a Mixed Economy: Class Struggle and State Policy in the Nicaraguan Transition," in *Transition and Development: Problems of Third World Socialism,* ed. Richard R. Fagen, Carmen Diana Deere, and José Luis Coraggio [New York: Monthly Review Press, 1986], pp. 179–181).

71. Xabier Gorostiaga, "Dilemmas of the Nicaraguan Revolution," in *The Future of Central America,* ed. Richard R. Fagen and Olga Pellicer (Stanford: Stanford University Press, 1983), p. 49; and Austin, Fox, and Krüger, "Role of the Revolutionary State," pp. 16–17. For an assessment of the problems of Nicaragua's transition to socialism, see Fagen, Deere, and Coraggio, eds., *Transition and Development.*

Table 7. Estimated structure of land tenure in 1990 and 2000

Ownership	Share of total farm land (%)		
	1981–82	1990	2000
Socialized sector	33.2	51.90	75.8
APP (state farms)	18.3	22.29	27.4
CAS (Sandinista Agric. Coop.)	1.3	11.75	25.1
CCS (Credit and Service Coop.)	13.6	17.85	23.3
Large private farms	12.0	9.36	6.0
Small and medium-size private farms	54.8	37.75	18.2

Source: MIDINRA, *Marco estratégico del desarrollo agropecuario*, vol. 1 (Managua: MIDINRA, 1983), p. 15.

871 acres (500 *manzanas*), and small and medium-size farmers (less than 871 acres) would see their holdings affected by land reform expropriations and partial cooperativization. State-owned agribusiness enterprises representing the Area of People's Property (APP), Sandinista Agricultural Cooperatives (CAS), where land, capital, and labor were organized collectively, and Credit and Services Cooperatives (CCS), which grouped independent peasants with some centralized procurement functions, would all increase their importance and would form what was called the new "socialized sector" (*sector socializado*). State enterprises were assigned the role of primary implementer of government policy and the center of profit and investment for the entire agricultural sector.[72]

According to MIDINRA strategists, state farms and cooperatives were projected to increase their combined share of the country's cultivated landholdings from 33 percent in 1981–82 to 76 percent in the year 2000, with large, medium-size, and small independent farmers decreasing their combined shares from 67 percent to 24 percent (see Table 7). In terms of production output, the socialized sector was expected to increase its share from 37 percent to 70 percent and the private sector to decrease from 63 percent to 30 percent (see Table 8).

72. MIDINRA, *Marco estratégico del desarrollo agropecuario*, vol. 1, pp. 10–16.

44 Hungry Dreams

Figure 2. State participation in the Nicaraguan food system: estimated share of output, 1983

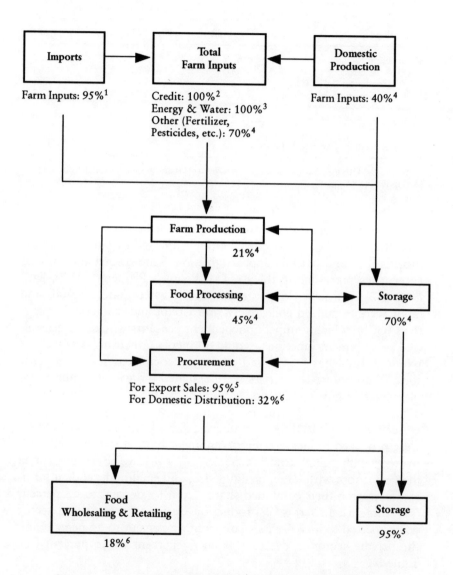

Sources: [1]Ministry of Foreign Trade (MICE), Enterprise of Foreign Inputs (ENIA); [2]Central Bank of Nicaragua (BCN); [3]Nicaraguan Energy Institute (INE); [4]Ministry of Agriculture and Land Reform (MIDINRA); [5]MICE; [6]Ministry of Internal Trade (MICOIN).

Table 8. Estimated share of output value by type of land tenure (%)

Ownership	Average 1980–82		1990		2000	
APP (state farms)	16.0		23.0		30.0	
CAS (Sandinista Agric. Coop.)	3.0	37.0	11.0	50.0	20.0	70.0
CCS (Credit and Services Coop.)	18.0		16.0		20.0	
Large private farms	14.0	63.0	10.0	50.0	5.0	30.0
Small and medium-size private farms	49.0		40.0		25.0	

Source: MIDINRA, *Marco estratégico del desarrollo agropecuario: Resumen ejecutivo* (Managua: MIDINRA, 1983), p. 13.

The Role of State Enterprises

With their establishment in 1979, state-owned enterprises rapidly became an all pervasive component of the national food system, having a significant presence in inputs, production, processing, and distribution (see Figure 2).[73] Banks were nationalized, and all credit and insurance operations came under the control of state-owned financial institutions. Almost all import and export channels were assigned to state trading companies, with the Central Bank having exclusive control over foreign exchange. Fertilizers, pesticides, agricultural machinery, seeds, fuel, and most other agricultural inputs were primarily or exclusively imported and distributed by state companies. In addition to state-owned farms and food-processing plants, the state also controlled basic grains procurement, storage, and distribution networks.

Being direct extensions of the government, these state-owned companies were viewed as a reliable policy instrument for generating loyal and rapid response to the new food policy objectives. Accordingly, they were assigned major food policy implementation tasks and primary responsibility for generating profit and foreign exchange, accumulating capital, and coordinating the growth of the domestic food system.[74] On the consumption side, the National Supply Company (ENABAS), a state agency, began administering

---

73. For an analysis of the role of the state in the Nicaraguan food system, see Austin, Fox, and Krüger, "Role of the Revolutionary State," pp. 15–40.
74. See, for example, Henry Ruiz, *El papel político del APP*; MIDINRA, *Marco estratégico del desarrollo agropecuario*, vol. 1, pp. 7–75; and MIDINRA, *Política agropecuaria en Nicaragua, 1981* (Managua: MIDINRA, 1981).

food price controls, consumption subsidies, procurement of basic grains, a network of People's Stores, and supplies to state-controlled supermarkets and Workplace Commissaries.

On the production side, most programs were coordinated or implemented by state enterprises. The Plan for Livestock Development was centered around the Chiltepe and Muy Muy-Matiguas state enterprises. The Plan for Vegetables Development was developed by the Valle de Sebaco and IFRUGALASA (the fruit and vegetable canning-bottling company) integrated state-owned processing plants. The main beneficiary of the Plan for Sugar Development was the Victoria de Julio state sugar mill. The National Food Program covering basic grain production came to be administered by PAN, a state agency reporting to MIDINRA. Sugar, cotton, coffee, and banana exports came under ENAZUCAR, ENAL, ENCAFE, and BANANIC, respectively, all state trading companies reporting to the Ministry of Foreign Trade (MICE). New investments in agriculture were channeled through the Financial Corporation (CORFIN) and later the National Investment Fund (FNI), both belonging to the National Financial System (SFN), which incorporated all state financial institutions. The portfolio of commercial loans to agriculture held by the formerly private Banco de America and Nicaraguan Bank came under the administration of the state-run National Development Bank (BND). The dissemination of improved farming technologies was assigned to a number of state agencies and to the state enterprises EMPROSEM (seeds), AGROMEC (farm machinery), PROAGRO (fertilizers, pesticides, artificial insemination), ETSA (airplane spraying), and SUMAGRO (inputs for agricultural processing).

The Implementation Record: Unexpected Difficulties

Summarizing the mixed performance of the state in the Nicaraguan food system during the first four years following the revolution, James Austin, Jonathan Fox, and Walter Krüger said, "Many difficulties were encountered and mistakes made and although many problems remain, significant production and consumption gains were achieved."[75] Their analysis of the food system also included a warning about major food policy challenges lurking ahead, such as

---

75. Austin, Fox, and Krüger, "Role of the Revolutionary State," p. 15.

effective macroeconomic management, the strengthening of state-owned enterprises' performance, the preservation of a pragmatic problem-solving orientation in the search for innovative ways to ensure economic and political pluralism, and the avoidance—in the face of an ongoing war—of strategic and structural distortions in the allocation of resources that would undermine recent food system gains.[76]

In 1985, the weight of past mistakes and difficulties appeared much greater. Fewer gains were recorded, and some of the major food policy challenges that had been predicted had become a powerful reality. The government's overall assessment, formulated in the 1985 Economic Plan in unusually blunt terms, warned that the economy had reached a watershed and expressed concern over the feasability of government policies and the survival of the revolutionary government.

> In 1985 the effects of the accumulated deterioration of the country's economic situation will continue to be felt, and the likelihood is that they will get worse.... The phenomena that demonstrate with more clarity the deterioration of the economic situation are the following: (a) A marked decrease in the supply of goods and services to the population, which becomes more acute starting in 1982, when the foreign exchange crisis begins to surface. (b) An increasing deterioration of installed plant capacity and transportation, due to the inability to ensure adequate replacement and maintenance of the equipment. This indicates growing inefficiency and a diminished capacity to produce goods and services. (c) An inflation that has been growing year after year, which affects in a fundamental way the income level of wage workers and discourages labor from joining activities in production and essential services, such as public health. (d) The flourishing of a speculative commercial sector which distorts the prices of products, removes huge quantities of cash from the National Financial System, attracts labor, and has come to represent a sort of parallel and parasitic economy which increases at the expense of what might be called the formal economy. (e) The accelerated decapitalization of the National Financial System, which is due to the foreign exchange losses, bad loans [made by state banks], and the fiscal deficit generated by the government's central administration. (f) A situation of chaos in the pricing system, which introduces irrationality in the allocation of products, discourages production, and promotes speculation and smuggling. (g) A drastic drop in average labor productivity, particularly in the agricultural sector, which causes cost increases and a decrease in production. All these phenomena must be controlled to avoid greater distortions in the national economy, which in a

---

76. Ibid., p. 35.

relatively short period of time can constitute a serious threat to the stability and the consolidation of the revolutionary power.[77]

The deterioration of the economy (see Table 9) had actually started in 1982, gradually progressed during 1983 and 1984, and was fully detected only in 1985, when a number of negative trends reached critical proportions. The economy took a sharp downturn, profoundly affecting Nicaragua's food system and calling into question important elements of the government's food policy.

On the consumption side, the purchasing power of wages had dropped by more than 80 percent between 1980 and 1985, although nominal wages during the same period tripled. Likewise, in 1985 the consumer price index was fourteen times higher than in 1980, and real per capita consumption was 52 percent of what it had been in 1980. This outcome was mostly due to accelerating inflation that, according to conservative government estimates, had increased from 50 percent in 1984 to 334 percent in 1985 and appeared to be pushing labor away from production into more remunerative informal trading activities, such as food distribution outside the formal marketing channels.

On the production side, gross domestic product decreased by 4 percent in real terms in 1985 and was still 26 percent below 1977, the last year that the government considered "normal," that is, unaffected by the social disturbances culminating in the 1979 revolution.[78] Meanwhile, the population in 1985 was estimated to have increased by 18 percent compared to 1980 and by 28 percent compared to 1977.[79] Export revenues in 1985 were at a record low of $301 million, down from $450 million in 1980 and a national average of $600 million in the second half of the 1970s. Physical output in the agricultural sector appeared to have fared better (Table 10),[80] but agricultural exports had decreased (Table 11) and mounting

---

77. SPP, *Plan económico 1985* (Managua: SPP, 1985), pp. 1–3.
78. Ministry of Planning, *Programa de reactivación económica en beneficio del pueblo, 1980* (Managua: MIPLAN, 1980), p. 17.
79. INEC, *Anuario estadístico de Nicaragua*.
80. There are numerous inconsistencies in the statistical data on Nicaraguan farm production provided by the various government institutions that collect them (primarily the National Institute of Statistics [INEC], the Central Bank of Nicaragua, the other institutions of the National Financial System, the Secretaría de Planificación y Presupuesto—that is, the former Ministry of Planning—the Ministry of Agriculture and Land Reform, the Ministry of Industry, the Ministry of Foreign Trade, and the Ministry of Internal Trade). Inconsistencies also exist between the data collected by different branches of the same institution (within the Ministry of Agriculture, for example, between the data collected by

Table 9. Basic economic indicators, 1980–1985

	1980	1981	1982	1983	1984	1985
Consumer price index	100	133.4	163	216.6	325.3	1,412.87
Inflation rate	35.30%	33.40%	22.19%	32.88%	50.18%	334.33%
Real GDP (1980 = 100)	20,798.8	21,914.3	21,735.4	22,701.1	22,368.9	21,468.4
GDP rate of growth	4.60%	5.36%	-0.82%	4.44%	-1.46%	-4.03%
Unemployment rate	37.20%	34.00%	32.90%	29.10%	26.20%	26.00%
Private consumption (1980 = 100)	17,165.9	14,737.9	12,842.6	12,119.6	11,494.6	10,581.8
Population (millions)	2.80	2.90	3.00	3.10	3.20	3.30
Per capita consumption (real)	6,130.68	5,082.03	4,280.87	3,909.55	3,592.06	3,206.61
Exports (US$ millions)	450.4	499.8	450.8	428.8	385.7	301.5
Imports (US$ millions)	802.9	922.5	723.5	778.1	799.6	878.2
Current account deficit	379.10	505.90	470.70	441.70	457.70	627.00
Average wage (nominal)	5,496	5,544	5,560	5,560	8,350	16,700
Average wage (real)	5,469	4,156	3,411	2,567	2,567	1,182

Source: Central Bank of Nicaragua.

Table 10. Farm output, 1977–1985

	1976–77	1977–78	1978–79	1979–80	1980–81	1981–82	1982–83	1983–84	1984–85
Area harvested (000 mz)									
Export[a]	475.0	493.8	456.0	279.8	367.2	362.6	358.4	395.8	392.3
Sesame	8.3	8.6	9.0	25.0	33.0	20.5	13.9	22.0	24.3
Bananas	3.3	3.5	3.5	3.6	4.2	3.9	3.8	3.8	3.8
Coffee	120.0	120.0	135.0	140.0	134.7	140.0	142.5	138.0	131.4
Sugarcane	59.6	57.4	59.3	55.7	59.3	64.5	68.0	65.0	65.0
Tobacco	0.8	0.9	1.0	0.9	1.3	1.0	1.1	0.9	1.2
Cotton	283.0	303.4	248.2	54.6	134.7	132.7	129.1	166.1	164.9
Domestic[b]	531.9	489.8	534.1	404.3	567.9	548.2	506.2	516.9	437.0
Rice	30.0	35.0	39.4	30.0	60.5	52.7	67.6	54.9	58.6
Beans	96.5	88.0	95.0	75.0	110.0	135.4	106.4	133.4	102.2
Corn	323.8	303.2	325.0	240.0	317.5	296.8	273.5	262.0	201.3
Sorghum	80.0	62.0	73.0	58.0	78.3	62.3	57.7	64.0	72.6
Tobacco	1.6	1.6	1.7	1.3	1.6	1.0	1.0	2.6	2.3
Total	1,006.9	983.6	990.1	684.1	935.1	910.8	864.6	912.7	829.3
Volume produced (000 cwt)									
Export[a]	62,326.5	62,214.9	66,778.7	52,302.5	57,375.8	67,510.3	66,255.2	70,107.9	5,777.3
Sesame	75.8	100.7	126.0	181.9	206.5	160.0	111.0	234.8	155.0
Bananas	3,448.5	3,533.3	3,774.1	2,720.0	2,682.1	3,360.0	3,099.0	2,604.0	2,820.0
Coffee	1,230.0	1,200.0	1,415.0	1,224.0	1,248.9	1,328.0	1,568.4	982.0	1,025.0
Sugarcane	54,992.0	54,681.8	58,968.4	47,675.3	51,556.9	61,250.9	59,691.0	64,167.7	52,671.5
Tobacco	23.2	25.9	29.0	27.2	34.9	24.0	28.2	18.3	27.5
Cotton	2,557.0	2,673.2	2,466.2	474.1	1,646.5	1,387.4	1,757.6	1,887.1	1,700.0
Domestic[b]	7,609.0	6,854.0	9,460.5	6,821.5	8,623.5	9,115.3	8,307.1	10,402.5	9,886.5
Rice	825.0	1,050.0	1,300.2	816.0	1,352.0	1,736.0	2,103.6	2,172.7	2,203.1
Beans	1,177.3	894.9	1,206.5	862.5	845.7	1,205.0	1,100.9	1,261.8	965.4
Corn	4,371.3	3,942.0	5,525.0	3,750.0	4,281.2	4,361.9	3,918.6	4,810.6	4,106.8
Sorghum	1,200.0	930.0	1,387.0	1,350.0	2,100.0	1,780.8	1,153.4	2,104.8	2,555.1
Tobacco	35.4	37.1	41.8	43.0	44.6	31.6	30.6	52.6	56.1
Total	69,935.5	69,068.9	76,239.2	59,124.0	65,999.3	76,625.6	74,562.3	80,540.4	15,664.0

Table 10. Continued

	1976–77	1977–78	1978–79	1979–80	1980–81	1981–82	1982–83	1983–84	1984–85
Yields (cwt/mz)									
Export[a]									
Sesame	131.2	126.0	146.4	186.9	156.3	186.2	184.9	177.1	14.7
	9.1	11.7	14.0	7.3	6.3	7.8	8.0	10.7	6.4
Bananas	1,045.0	1,009.5	1,078.3	755.6	638.6	861.5	815.5	743.2	742.1
Coffee	10.3	10.0	10.5	8.7	9.3	9.5	11.0	7.1	8.2
Sugarcane	922.7	952.6	994.4	855.9	869.4	946.6	877.8	987.2	810.3
Tobacco	29.0	28.8	29.0	30.2	26.8	24.0	25.6	20.3	22.9
Cotton	9.0	8.8	9.9	8.7	12.2	10.5	13.6	11.3	10.3
Domestic[b]									
Rice	14.3	14.0	17.7	16.9	15.2	16.6	16.4	20.1	22.6
Beans	27.5	30.0	33.0	27.2	22.3	32.9	31.1	39.6	37.6
Corn	12.2	10.2	12.7	11.5	7.7	8.9	10.3	9.5	9.4
Sorghum	13.5	13.0	17.0	15.6	13.5	14.7	14.3	18.4	20.4
Tobacco	15.0	15.0	19.0	23.3	20.8	28.6	20.0	32.9	35.2
	22.1	23.2	24.6	33.1	27.9	31.6	30.6	20.2	24.4
Total	69.5	70.2	77.0	86.4	70.6	84.1	86.2	88.2	18.9
Livestock									
Beef (million lbs)	125.0	139.1	116.0	103.7	78.1	92.2	98.3	91.7	98.5
Milk (million gals)	118.5	119.2	99.0	39.8	39.6	38.3	26.6	40.7	42.4
Poultry (million lbs)			23.4	18.9	24.5	29.2	28.5	25.9	22.6
Eggs (million dozens)			20.8	28.8	31.8	42.5	42.3	40.1	20.9
Pork (million lbs)			18.0	18.6	21.5	25.1	23.5	23.7	19.5

Source: Ministry of Planning.
Note: mz = manzana (1.7 acres); cwt = hundredweight (100 pounds).
[a]Production primarily for export markets.
[b]Production primarily for domestic consumption.

production and distribution problems were reported in the milk, livestock, cotton, cooking-oil, sorghum, rice, and animal-feed commodity systems.[81]

Food subsidies had increased dramatically and came to be viewed as unaffordable.[82] Agricultural credit faced the same challenge, and the magnitude of bad loans was threatening the economic viability of the banking system.[83] Lack of foreign exchange was creating a corresponding shortage in the full range of agricultural inputs, affecting food production and the ability to increase exports.[84] Furthermore a variety of interrelated shortages at various stages of the food system was influencing food supply and distribution.[85] As a result, the objectives of food self-sufficiency and food security came to be regarded as likely to be achieved in a more distant future than originally expected.

Government officials and other analysts attributed this economic deterioration to various causes. Prominent in the list was, obviously, the ongoing war, with its multiple direct and indirect effects.[86] Other reasons cited were the U.S. trade embargo of 1985, falling international prices for agricultural commodities, worsening terms of trade, the gradual disintegration of the Central American Common Market, and bad weather.[87] Among internal and seemingly more controllable causes, however, attention was increasingly focused on the performance of state-owned agribusiness enterprises, whose operations greatly affected the functioning of the food system and the state of the economy as a whole. Government concerns centered around four major areas: the state enterprises' persistent financial

---

CIERA, PAN, the various general and regional directorates, the departments of planning, auditing, investment, etc.). Among possible reasons for the discrepancies are (1) the fact that farm production is recorded by MIDINRA according to the agricultural cycle, whereas national budget statistics are recorded according to the calendar year; (2) confusion between actual, estimated, projected, and budgeted data; (3) differences in units of measurement (sometimes metric, sometimes nonmetric); and (4) changes in the definition of the product (for example, fluid milk alternatively as the milk that comes out of the cow or, more restrictively, the input of pasteurizing plants).

81. SPP, *Plan económico 1985*.
82. CIERA/PAN/CIDA, *Directorio de políticas alimentarias*, pp. 67–73.
83. Interviews with officers of the National Financial System (SFN), 1986 and 1987.
84. SPP, *Plan económico 1985*.
85. Ibid.
86. See, for example, the estimate of the government's adviser E. V. K. Fitzgerald in "Economic Cost to Nicaragua of U.S. Aggression," pp. 195–216.
87. For a critical review of these sources, see Francisco Mayorga, *Nicaragua: Trayectoria económica, 1980–1984—Algunas apreciaciones* (Managua, 1985); Mayorga, "Nicaraguan Economic Experience," and José Luis Medal, *Políticas de estabilización y ajuste estructural en Nicaragua (1980–1986)* (San José: CSUCA, 1987).

Table 11. Farm exports, 1977–1985

Product	1977	1978	1979	1980	1981	1982	1983	1984	1985
Agricultural products									
Coffee									
Value ($ million)	199.8	199.6	158.5	165.7	136.8	124.0	153.7	122.4	122.6
Volume (000 cwt)	1,078.0	1,188.0	1,204.0	1,000.0	1,132.0	1,012.0	1,423.0	891.5	898.2
Price ($/cwt)	184.8	168.0	131.6	165.7	120.9	122.5	108.5	137.3	136.5
Cotton									
Value ($ million)	150.6	140.9	135.7	30.4	123.4	87.2	109.5	133.8	91.0
Volume (000 cwt)	2,531.0	2,804.0	2,427.0	427.0	1,627.0	1,351.0	1,726.0	1,809.4	1,458.1
Price ($/cwt)	59.5	50.3	54.9	71.2	75.9	64.5	63.5	74.0	62.4
Sugar									
Value ($ million)	27.8	19.6	19.6	20.5	51.0	36.4	34.4	20.9	7.4
Volume (000 cwt)	2,157.0	2,126.0	1,974.0	1,348.0	2,232.0	2,066.0	2,376.0	2,200.0	1,374.3
Price ($/cwt)	12.9	9.2	9.9	15.2	22.9	17.6	14.5	9.4	5.5
Bananas									
Value ($ million)	4.5	4.8	6.4	8.4	20.9	9.8	14.8	11.9	14.9
Volume (000 cwt)	5,677.0	6,012.0	5,637.0	5,694.0	4,907.0	2,283.0	4,288.0	4,166.0	4,387.0
Price ($/box)	0.8	0.8	1.1	1.5	4.3	4.3	3.5	2.9	3.4
Sesame									
Value ($ million)	1.8	3.4	3.2	6.3	8.1	5.8	5.8	5.9	5.5
Volume (000 cwt)	59.0	111.0	106.0	117.0	118.0	108.0	100.0	147.1	125.4
Price ($/cwt)	29.8	30.7	29.7	53.7	43.6	53.9	58.4	40.2	44.1
Tobacco									
Value ($ million)	4.5	3.0	3.0	1.4	3.7	4.4	3.5	4.2	
Volume (000 kg)	1.5	1.1	1.0	0.4	1.0	1.0	0.8		
Price ($/kg)	3.0	2.7	3.1	3.4	3.9	4.4	4.5		
Corn									
Value ($ million)	0.5	0.2	0.0	0.0					
Volume (000cwt)	0.0	0.0	0.0						
Price ($/cwt)	66.9	56.0	12.0						

Table 11. Continued

Product	1977	1978	1979	1980	1981	1982	1983	1984	1985
Beans									
Value ($ million)	0.0	0.4	0.9	0.0	2.0	0.7	1.0		
Volume (000 cwt)	0.0	0.0	0.1		0.1	0.0	0.0		
Price ($/cwt)	10.0	18.9	10.3		38.8	38.5	31.7		
Lumber									
Value ($ million)	0.4	0.5	1.0	0.5	0.1	0.1	0.1	0.0	
Volume (000 ft)	1.6	1.6	1.4	2.0	0.1	0.4	0.0		
Price ($/ft)	0.2	0.3	0.7	0.2	0.8	0.3	0.3		
Livestock									
Beef									
Value ($ million)	37.3	67.7	93.5	58.6	23.2	33.8	31.4	17.6	12.6
Volume (000 cwt)	58.1	74.9	78.3	45.1	20.9	32.0	31.3	19.8	13.5
Price ($/cwt)	0.6	0.9	1.2	1.3	1.1	1.1	1.0	0.9	0.9
Hides									
Value ($ million)	3.8	5.6	3.3	2.5	1.5	0.4	2.4		
Volume (000 cwt)	0.6	1.0	0.8	0.3	0.1	0.0	0.2		
Price ($/cwt)	6.0	5.8	3.9	9.8	15.1	15.8	12.1		
Lobster and shrimp									
Value ($ million)	22.0	14.7	21.7	26.8	19.7	21.7	16.8	12.6	12.3
Volume	11,982.0	9,325.0	8,301.0	7,549.0	5,374.0	4,144.0	2,878.0	2,264.0	2,400.0
Price ($/cwt)	1.8	1.6	2.6	3.5	3.7	5.2	5.8	5.6	5.1
Total farm exports ($ million)	452.0	460.4	446.8	321.1	390.4	324.3	373.4	329.3	265.7

Source: Central Bank of Nicaragua.
Note: $ = U.S. dollar.

losses, their inability to service their heavy and growing debt, the underutilization of their food-processing plants and equipment, and an unsatisfactory and deteriorating level of their labor productivity.

These areas of unsatisfactory performance were absorbing a growing amount of increasingly scarce national resources, were thought to be partly responsible for the overall deterioration of the economy, and seemed to have contributed to the development of an ever more distorted food system, where—according to government reports—chaos in the pricing system, inflation, speculation in food distribution, smuggling, and distribution shortages were the more visible symptoms.[88] In the next four chapters, I analyze these performance difficulties and their impact on the food system within the broader context of macro price policy.

88. SPP, *Plan económico 1985*, pp. 1–13.

3

Foreign Exchange

All enterprises, especially state-owned enterprises, must be profitable and must contribute to the national budget.
—1981 National Economic Plan

During the past five years we have been losing a lot of money. Had the foreign exchange rate not been so distorted, we would have been profitable. But the state does not seem to understand that we are not getting the benefits they think they are giving us.
—Director, state-owned export farm, 1984

The dollar/cordoba exchange rate has been systematically overvalued. . . . Prices set by the government to farm producers have not been an incentive to production. Often they have been lower than production costs.
—1985 National Economic Plan

With the swift expropriation of Somoza's assets as well as those of his close allies, along with the nationalization of the banks and agricultural export marketing channels immediately following the 1979 revolution, the new government felt it had acquired a substantial portion of the agroexport "money machine" that had been responsible for past economic growth and for the accumulation of large profits in the hands of a small oligarchy.[1] This acquisition appeared to serve food policy well, since it placed state-owned agribusiness enterprises—and by extension, the government—in control of a sizable portion of the nation's traditional source of profits and foreign exchange. It also provided the means by which state enterprises could assume a leading role in building the "New Economy,"[2]

---

1. See, for example, Ministry of Planning, *Bases para la discusión de una estrategia económica sandinista* (Managua: MIPLAN, 1979), p. 62.
2. For a general treatment of the question of the transition to socialism and to a "New

whose structure was to be based on a more equitable and diversified food system.³

State-owned enterprises, however, did not fulfill the government's profit expectations. On the contrary, between 1979 and 1985 they reported substantial losses, becoming an increasing financial burden to the state. The losses undermined their ability to sustain the level of investment in the agricultural sector needed to meet food self-sufficiency objectives.⁴ The financial difficulties of state enterprises also motivated these firms to emulate the private sector by diversifying, whenever possible, away from agroexport into more profitable activities in the domestic market, further damaging the country's ability to generate much needed foreign exchange.⁵

A significant part of the losses incurred by state-owned agribusiness enterprises can be related to macroeconomic prices inconsistent with food policy objectives. This chapter shows how foreign exchange rates were set at a level that worked against the agricultural export sector, where most state enterprises operated (see Appendix). A complex pricing system, made up of multiple exchange rates, export incentives, tax exemptions, and guaranteed "cost-plus" farm-gate prices divorced from international market prices, effectively hid the overvaluation of the national currency and thus contributed to the perpetuation of disincentives in the agricultural export sector.

One result of this situation was that many of the traditionally export-oriented private farmers, particularly cotton growers and cattle ranchers, curtailed their export operations or switched them to more remunerative activities in the domestic market. This behavior was responsible for much of the drop in export volume that marked the period 1980–85 and for a corresponding fall in foreign exchange revenues (see Table 11). By 1985 the export volume of cotton, cof-

---

Economy," with an emphasis on the Nicaraguan experience, see Richard R. Fagen, Carmen Diana Deere, and José Luis Coraggio, eds., *Transition and Development, Problems of Third World Socialism* (New York: Monthly Review Press, 1986). See also E. V. K. Fitzgerald, "The Economics of the Revolution," in *Nicaragua in Revolution*, ed. Thomas W. Walker (New York: Praeger, 1982), pp. 203–221.

3. Ministry of Planning, *Estrategia económica sandinista*, pp. 18–29.

4. See, for example, Alejandro Argüello, Edwin Croes, and Nanno Kleiterp, *Nicaragua: Acumulación y transformación, inversiones, 1979–1985* (Managua, 1986), pp. 34–35. A description of the tensions in the national investment program after 1981 caused by insufficient self-financing is given in the 1987 Economic Plan, Chapter 7, which is dedicated to investment policy (SPP, *Plan económico 1987* [Managua: SPP, 1987], pp. 158–167).

5. This pattern, which will be mentioned in other sections of this chapter, is discussed in greater detail in Chapter 4 in conjunction with domestic food price policy.

fee, and beef, the three leading generators of foreign exchange, had fallen by 42 percent, 17 percent and 77 percent respectively, compared with 1977, a year the government considered "normal."[6] Their combined export value during the same period fell by 42 percent. The resulting lack of foreign exchange, first noticed in 1980, contributed to a deterioration in the nation's ability to import farm and food-processing inputs crucial to the maintenance and growth of the food system. Shortages of critical items such as fertilizers, pesticides, and farm equipment led to further reductions in farm output, which in turn decreased exports and foreign exchange receipts, creating the premises for a vicious downward economic spiral.[7] Most export-oriented state enterprises, loyally implementing the government's desire to generate foreign exchange, continued to produce for the export market, but at the expense of their profitability. Eventually, many of them had to reduce their export operations, too, in order to survive financially.[8] Nevertheless, whether through output reduction or financial losses, the food system's long-term growth capability was impaired.[9]

6. Ministry of Planning, *Programa de reactivación económica en beneficio del pueblo, 1980* (Managua: MIPLAN, 1980), p. 17. For Nicaraguan production and export data in both physical and monetary terms during the period 1977–85, see Tables 10 and 11.

7. An internal government report written in July 1980 estimated that "the present trend of the economy indicates . . . a collapse of the balance of payments in 1982. . . . The principal tension in this first stage of the transition consists in the insufficient expansion of production." The need to recover the prerevolutionary levels of export production is also emphasized in the 1981 National Economic Plan, which set as an objective for that year to increase exports by 45 percent (Ministry of Planning, *Programa económico de austeridad y eficiencia 1981* [Managua: MIPLAN, 1981], p. 161). Exports that year actually increased only 11 percent and decreased thereafter. In 1985 they were 40 percent lower than in 1981.

For an analysis of the foreign exchange shortage and its negative impact on the food system, see, for example, BCN, *Plan de fomento a las exportaciones* (Managua: BCN, 1981), which states that "an economic situation like the one of 1981 is not sustainable in 1982 or beyond. The social and political cost of a deep economic recession would be extremely high. . . . Insofar that we will be unable in 1982 to reactivate production, investment, and exports, the deep economic recession will be the order of the day because of the worsening of the foreign exchange shortage which will not allow the country to import the absolutely indispensable" (p. 6). See also SPP, *Plan económico 1985* (Managua: SPP, 1985), pp. 86–98.

8. This point will be analyzed in more detail in other sections of this chapter, as well as in Chapter 4.

9. This outcome highlighted the interdependence of export and domestic agriculture, which some of the earlier studies had underestimated (see, for example, Peter Marchetti, S. J., "Prólogo a la edición en Español," in Solon Barraclough, *Un análisis preliminar del sistema alimentario en Nicaragua* [Geneva: UNRISD, 1984], p. xii; and n. 5 of Chapter 2, this book).

For the argument that in small, trade-dependent Third World economies the external trade sector, rather than peasant food agriculture, is the main source of surplus for accumulation and that foreign exchange, rather than cheap wage goods, is the key economic constraint, see George Irvin, *Nicaragua: Establishing the State as the Centre of*

State Enterprises as Sources of Profit: Goal and Performance

The transition to a "New Economy" called for reorganization of the national food system in accordance with the objectives and strategy delineated in the previous chapter. At the core of this reorganization was the establishment of the state as the center of accumulation.[10] Accumulation was defined as the generation of surpluses in production and in the means of production, over and above the existing productive base.[11] Three sources of accumulation were identified: the profits generated by state-owned enterprises through their own productive activities, the transfer to the state of profits generated by the private sector, and international loans.[12] It was difficult to predict how effective the state would be in appropriating the profits of the private sector. Furthermore, foreign debt was expected to be a source of deficit rather than surplus in the long run. Therefore, the government deemed it important to be able to rely on an autonomous profit source for the implementation of its new farm and agroindustrial policies.[13] Hence the insistence that each state-owned enterprise be profitable and that the Area of People's Property, the APP, be the main source of state surplus so as to assure the economic viability of the various food strategy components.[14] State enterprises were to represent the pillar of the new accumulation process and, consequently, the essence of the new Sandinista economy. Shortly

---

Accumulation (The Hague: Institute of Social Studies, 1982); E. V. K. Fitzgerald, "The Problem of Balance in the Peripheral Socialist Economy: A Conceptual Note," in *World Development* 13 (January 1985), pp. 5–14; and E. V. K. Fitzgerald, "Planned Accumulation and Income Distribution in the Small Peripheral Economy," in *Towards an Alternative for Central America and the Caribbean*, ed. George Irvin and Xabier Gorostiaga (London: Allen and Unwin, 1985).

10. See, for example, Ministry of Planning, *Estrategia económica sandinista*; Henry Ruiz, *El papel político del APP en la nueva economía sandinista* (Managua: FSLN, 1980); and Nicaraguan Institute of Public Administration (INAP), *Análisis de la gestión de las empresas públicas en Nicaragua* (Managua: INAP, 1981). See also Irvin, *Nicaragua: Establishing the State*; E. V. K. Fitzgerald, "Notes on the Analysis of the Small Underdeveloped Economy in Transition," in *Transition and Development*, ed. Fagen, Deere, and Coraggio, pp. 28–53; and John Weeks, "The Mixed Economy in Nicaragua: The Economic Battlefield," and David F. Ruccio, "The State and Planning in Nicaragua," in *The Political Economy of Revolutionary Nicaragua*, ed. Rose J. Spalding (Boston: Allen and Unwin, 1987), pp. 43–60 and 61–82. See also the informative thesis of Gustavo E. Melazzi, "Los problemas de la transición al socialismo: El caso de Nicaragua 1979–1981" (Ph.D. diss., Universidad Nacional Autónoma de México, 1985).

11. Ministry of Planning, *Estrategia económica sandinista*, p. 59.

12. Ibid., p. 61.

13. Ibid., p. 61; and Ruiz, *El papel político del APP*.

14. Ruiz, *El papel político del APP*, pp. 8–18.

after the revolution, Henry Ruiz, minister of planning, explained in the following terms the role to be played by the state-owned agribusiness enterprises making up the APP:

> Understanding the political role of the APP . . . amounts to understanding the essence of the change that is sought in the new Sandinista economy, its form, and the content of its functioning. . . . The new Sandinista economy aims as its central objective at the increasing satisfaction of the needs of our people traditionally exploited by the oligarchy and imperialism. . . . This implies a profound change; it implies that the economy ceases to be at the service of those [social] classes which, by virtue of their private ownership of the means of production, are able to appropriate themselves of a portion of the value produced by the workers, and to use it in an unbridled search for new sources of profit. The traditional economy, the one that made Somoza and its regime, had as its end the enrichment of the few and the impoverishment of the many. . . . The new Sandinista economy, on the contrary, will focus its effort in satisfying the increasing needs of the [social] classes that create the wealth of a nation: its workers. . . . This increasing satisfaction of social needs can only occur within a framework of accumulation. . . . All this process will be strengthened insofar as the support of the working class, and its spirit of sacrifice, will allow the establishment of the economic base which would make possible society's conscious regulation of the economy, which implies the consolidation and the advance of the APP. This is the essence of the new accumulation, and hence of the Sandinista economy. . . . Insofar as the APP aims at the satisfaction of the basic needs of the population, . . . it must generate social investment funds that allow an autonomous process of accumulation. . . . That is, [in this way] the process of surplus generation is merged with a mechanism that uses it in a manner consistent with the objectives of the Sandinista Popular Revolution, rather than for individual enrichment. . . . The creation of the APP raises the possibility . . . that the means of production be used in a manner consistent with the objectives of society, thereby eliminating man's voracity over his fellows. . . . To handle the APP in an irresponsible way would amount, in dynamic terms, to leaving adrift the riches that were snatched away from the Somoza regime. . . . It would amount to losing the great regulatory capability that is implied in [state] ownership of more than 30 percent of the means of production and services. . . . It would amount to transforming the July 19 [1979] victory over Somoza into Pyrrhic victory. . . . The APP, in other words, is the central pillar, the most dynamic pulley of the economic and social transformations of the revolution. . . . The economic surpluses that originate in the APP . . . will strengthen the APP's consolidation and expansion, allowing our people to share wealth, and not the opposite, as it happens today.[15]

15. Ibid., pp. 1–24.

The APP did not generate surpluses, however. A financial survey made by MIDINRA in 1981 already indicated that out of forty-nine state enterprises for which updated financial statements existed, thirty-eight were operating at a loss. Of those thirty-eight enterprises, thirty were export–oriented, producing primarily cotton, beef, coffee, and sugar. Moreover, each of the nine consolidated financial statements of the state enterprises, one national and eight regional, showed net losses.[16] The losses prompted the government's decision in 1982 to inject additional state funds into the state enterprises, raising their equity as a percentage of total assets from approximately 26 percent to 42 percent.[17] The stated rationale for this transaction, which included a partial cancellation and restructuring of the state enterprises' debt, was "that in order to evaluate the economic performance of the same [state-owned enterprises], they must be endowed with an adequate financial structure."[18] But the new financial structure resulting from the so-called *saneamiento financiero* (financial bailout) soon proved inadequate in the face of continued losses. A new financial survey in 1985 prompted the negotiation of yet another bailout. The new survey indicated that between 1982 and 1985, of forty-five enterprises for which up-to-date financial statements existed, thirty continued to operate at a loss and were more indebted than before the 1982 bailout. Of those thirty enterprises, twenty-six were export-oriented, producing, as in 1981, mostly cotton, beef, coffee, and sugar. Every regional consolidated statement reported financial losses, while the total indebtedness of the state enterprises had trebled.[19]

In 1980 the government's Economic Reactivation Plan stated, "The characteristics of the [economic] reactivation process guarantee the obtainment of profits from the state sector. . . . Somewhere along this reactivation the critical question will become the need to use those profits productively."[20] How unexpected APP's financial

16. See MIDINRA, *Estados financieros: Junio 1981* (Managua: MIDINRA, 1981); and MIDINRA, *Balances generales consolidados de las regiones al 30 de junio de 1981* (Managua: MIDINRA, 1981).
17. See MIDINRA, *Saneamiento financiero de las empresas de reforma agraria*, 4 vols. (Managua: MIDINRA, 1982).
18. See *Acuerdo: La Junta de Gobierno de Reconstrucción Nacional de la República de Nicaragua* (Managua: November 2, 1982), p. 1; and *Acuerdo: El Ministerio de Desarrollo Agropecuario y Reforma Agraria* (Managua: March 1983), p. 1.
19. See MIDINRA, *Balances generales consolidados al 31 de marzo de 1985* (Managua: MIDINRA, 1985). For a more extended treatment of the question of state enterprises' indebtedness, see Chapter 6.
20. See Ministry of Planning, *Programa de reactivación económica en beneficio del pueblo*, p. 25.

losses were is perhaps best illustrated by a proposal made by the Ministry of Planning (MIPLAN) in 1980, which suggested that a special State Accumulation Fund be created under the administration of the Central Bank and the Ministry of Finance for the purpose of centralizing the profits of state enterprises and the planning of future investments.[21] The proposal was dropped when, in the words of a Central Bank official, "somewhere in 1982 it became obvious that the problem was not how to capture and allocate state enterprise profits, but rather how to reduce their losses and repay their mounting debts to the banking system."[22]

Accordingly, in 1982 an extensive program aimed at improving the efficiency of state enterprises was initiated. Some projects were coordinated by MIDINRA's Entrepreneurial Organization and Administration Division (DOGE) under the supervision of a vice-minister, while others were of an interinstitutional nature. Goals included the design and implementation of improved administrative systems, particularly in the area of planning, budgeting, and control; clearer specification of state enterprise objectives; the streamlining of many enterprises through mergers, divestitures, and more rational allocation of land and farm equipment; partial decentralization of decision making; the introduction of new accountability and motivation systems; the analysis of entire agribusiness commodity systems aimed at better overall system coordination; and a comprehensive training program in basic administration skills for all state-owned enterprise managers.[23]

---

21. For an analysis of this proposal and its background, see IBRD, *Nicaragua: The Challenge of Reconstruction* (Washington, D.C.: World Bank, 1981), p. 42; and MIDINRA, *Política agropecuaria en Nicaragua, 1981* (Managua: MIDINRA, 1981), pp. 26–30.

22. Interview, August 1987. Earlier indications that APP's financial performance was at variance with government expectations do not seem to have been sufficiently noticed. See, for example, CIERA, *Ministerio de Desarrollo Agropecuario y el APP agropecuario* (Managua: MIDINRA, 1980), pp. 4–7.

23. For a description of the DOGE project, see, for example, DOGE, *Orígenes, desarrollo y perspectiva de la planificación y control financiero en las empresas del APP adscritas al MIDINRA* (Managua: MIDINRA, 1983); DOGE, *Gestión de empresas de reforma agraria* (Managua: MIDINRA, 1984); DOGE, *Sistema de dirección de empresas* (Managua: MIDINRA, 1984); DOGE, *Actividades a desarrollar por la división de organización y gestión empresarial* (Managua: MIDINRA, 1985); DOGE, *Notas sobre DOGE* (Managua: MIDINRA, 1985); and DOGE, *Plan de trabajo para el fortalecimiento del sistema de empresas estatales en el sector agropecuario* (Managua: MIDINRA, 1986). For a description of the SORINADE project, see *Informe a la JGRN sobre el avance del proyecto: Sistema de organización e información administrativa del estado (SORINADE)* (Managua: MIPLAN, INAP, Bulgarian Consulting, 1983); and *Proyecto SORINADE programa, 1984/1985* (Managua: MIPLAN, INAP, Bulgarian Consulting, 1984). For an analysis of APP's reorganization, see Walter Krüger and James E. Austin, *Organization*

In spite of significant improvements in these areas, however,[24] state enterprises continued to show negative financial results.[25] Accordingly, more visible steps, such as the use of the mass media, were initiated to remind state administrators and the public at large of how important it was for the economy that state enterprises be profitable.[26] This effort focused almost exclusively on ways to improve the internal efficiency of state enterprises.[27] Meanwhile, negotiations for the second bailout began, partly as a response to the mounting tensions between MIDINRA and the National Financial System (SFN), whose managers were concerned that APP's growing unpaid debt would threaten the financial stability of the banks.[28]

Financial Losses in the Export Sector: The Case of a State Cotton Farm

The Challenge of Export Agriculture

Cotton presented a number of special food policy challenges to postrevolutionary policymakers seeking to respond to the complex issues raised by the diagnosis of the food system.[29] On the one hand, it represented perhaps the most visible export success of the previous decades. Since 1950 it had been Nicaragua's fastest growing agricultural product and the major source of foreign exchange, cooking oil, animal feed, and seasonal employment.[30] Starting from an almost

---

and Control of Agricultural State-owned Enterprises: The Case of Nicaragua (Boston: Harvard Business School, 1983).

24. See, for example, MIDINRA, *Evaluación al grado de implementación del sistema unitario de control administrativo* (Managua: MIDINRA, 1984). Improvements in the area of planning, budgeting, and control will be discussed in more detail in Chapter 6.

25. MIDINRA, *Balances generales consolidados al 31 de marzo de 1985*.

26. See, for example, Jaime Wheelock's speech in the March 7, 1985, issue of *Barricada*.

27. See, for example, MIDINRA, *Evaluación de las empresas adscritas a MIDINRA* (Managua: MIDINRA, 1985); MIDINRA, *Empresas adscritas al MIDINRA: Reflexiones sobre la gestión actual y sugerencias para mejorarla* (Managua: MIDINRA, 1985); MIDINRA, *Revisión integral de las empresas estatales agropecuarias y fortalecimiento de sus sistemas de dirección y gestión* (Managua: MIDINRA, 1985).

28. Interviews with officers of the Central Bank and the National Financial System, 1985 and 1986. See also BCN, *El Area Propiedad del Pueblo, su relación con el sistema financiero nacional: Problemas* (Managua: BCN, 1984). From the perspective of the APP, the issue of state-owned enterprises' financial difficulties and their relationship with the banking system is partly covered in MIDINRA, *Problemática de las empresas del sector agropecuario adscritas al MIDINRA* (Managua: MIDINRA, 1985). See also Chap-ter 6.

29. See Chapter 2.

30. See, for example, Pedro Belli, "An Inquiry Concerning the Growth of Cotton

64 Hungry Dreams

insignificant production base in the late 1940s, in just one decade Nicaragua had become an important supplier to the international market and the leading cotton grower in Central America. Nicaraguan cotton yields ranked among the world's highest and were frequently higher than those of U.S. producers.[31] By 1978 the ramifications of the cotton agribusiness commodity system[32] pervaded the entire domestic food economy and played a prominent role in the development of the country.

Additionally, particular circumstances of the revolution justified a special focus on this crop. In 1979 the cotton-planting season had coincided with the peak of revolutionary activity, and the acreage devoted to the white fiber had fallen more than that of any other crop (see Table 10). The full impact of this decrease in acreage was felt only in early 1980, when it became apparent that cotton output in the crop year 1979–80 would be one-fifth that of the preceding years and that foreign exchange receipts from cotton exports—the first that the new government was in a position to collect after assuming power—had fallen by $100 million, despite higher prices (see Table 11).[33] Thus the new government had at least two reasons for wanting to increase cotton production. First was the fiber's strategic role in the domestic food system, a role that no other commodity could fill in the immediate future.[34] Second was an incipient

---

Farming in Nicaragua" (Ph.D. diss., University of California, Berkeley, 1968); David R. Radell, "Historical Geography of Western Nicaragua 1519-1965" (Ph.D. diss., University of California, Berkeley, 1969); Orlando Núñez Soto, *El somocismo y el modelo capitalista agroexportador* (Managua: UNAN, 1980); OEDEC, *El algodón en Nicaragua* (Managua: OEDEC, 1979); and Robert G. Williams, *Export Agriculture and the Crisis in Central America* (Chapel Hill: University of South Carolina Press, 1986).

31. *Historical Geography*, p. 240; and Belli, "Cotton Farming in Nicaragua," pp. 57–64.

32. The term "agribusiness commodity system" refers to a whole conceptual approach to the business of agriculture, as developed at the Harvard Business School by John H. Davis and Ray A. Goldberg in *A Concept of Agribusiness* (Boston: Harvard Business School, 1957) and further refined by Goldberg in *Agribusiness Coordination* (Boston: Harvard Business School, 1968). The agribusiness commodity system "encompasses all the participants involved in the production, processing, and marketing of a single farm product. Such a system includes farm suppliers, farmers, storage operators, processors, wholesalers, and retailers involved in commodity flow from initial inputs to the final consumer. It also includes all the institutions which affect and coordinate the successive stages of a commodity flow, such as the government, futures markets, and trade associations" (Goldberg, *Agribusiness Coordination*, p. 3).

33. See Ministry of Foreign Trade, *Evaluación de las ventas de la empresa nicaragüense de algodón (ENAL)* (Managua: MICE, 1982); MIDINRA, *Marco estratégico del desarrollo agropecuario*, vol. 4 (Managua: MIDINRA, 1983), p. 26; and Ministry of Planning, *Programa de reactivación económica, 1980*, p. 58.

34. See, for example, MIDINRA, *Marco estratégico del desarrollo agropecuario*, vol. 4

foreign exchange shortage that, if allowed to grow, could seriously affect the implementation of major food policy goals. Here again, as far as the future was concerned, the potential contribution of cotton was considered critical.[35]

Accordingly, it was argued that the function served by cotton in the past could not be underestimated or dismissed altogether without risking potentially serious economic consequences, at least in the short and medium term. It was not surprising, therefore, that as the 1980 cotton-planting season approached, the government was eager to recover prerevolutionary acreage levels and even to surpass them in the long run.[36] On the other hand, the new government fully subscribed to the view that Nicaragua's inherited hunger and malnutrition problems were partly attributable to the past cotton boom. As explained in the previous chapter, the expansion of cotton was thought to have been partly responsible for the displacement of basic grains production to less fertile lands in the interior, for a rise in the number of landless peasants, and for the perpetuation of rural poverty.[37] As a result, any decision concerning the future of cotton cultivation called for a particularly careful assessment of its consistency with food policy objectives.[38]

The essence of the challenge turned around a basic question: how could the food system take advantage of the potential benefits of cotton cultivation, while at the same time ensuring that cotton would not be grown, as in the past, at the expense of the basic needs of the population? The answer to this question was to be found in a restructuring of the cotton agribusiness commodity system to include new system participants and different pricing and contractual

---

(Managua: MIDINRA, 1983). The Instituto Histórico Centroamericano writes: "Why did the revolutionary government decide to give priority to the production of coffee and cotton? The sad truth is that, in the short run, there was no alternative. It was essential for Nicaragua to increase its generation of foreign exchange" (ENVIO, no. 29 [November 1983]: 3c).

35. See, for example, BCN, *Plan de fomento a las exportaciones*; and MIDINRA, *Marco estratégico del desarrollo agropecuario*, vol. 4, p. 24.

36. MIDINRA, *Marco estratégico del desarrollo agropecuario*, vol. 4, pp. 11, 24–27; and CIERA/PAN/CIDA, *Informe final del proyecto de estrategia alimentaria*, vol. 2 (Managua: MIDINRA, 1984), p. 7.

37. See Instituto Histórico Centroamericano, *ENVIO*, pp. 2c–4c; and Chapter 2, "The Implementation Record."

38. See, for example, Ministry of Planning, *Cultivo de algodón: Método y criterios para determinar el nivel de producción y el precio al productor* (Managua: MIPLAN, 1982); Instituto Histórico Centroamericano, *ENVIO*, pp. 2c–4c; and Solon Barraclough, *A Preliminary Analysis of the Nicaraguan Food System* (Geneva: UNRISD, 1982), p. 61.

arrangements.[39] From the government's perspective, the renewed expansion of cotton cultivation was fundamentally different from the prerevolutionary growth strategy, precisely because of the changes in the structure of the cotton agribusiness commodity system introduced after 1979.

Although an overall assessment of these changes is beyond the purpose of this book, several contributions in that direction have been made elsewhere.[40] A brief review of these innovations sheds light on the role and performance of state-owned export enterprises as new system participants.

Designing a Structure Consistent with Food Policy

Most structural changes in the cotton commodity system fell into two broad categories. Some changes were institutional and aimed at bringing greater direct state participation into the daily operations of the system. These changes involved the creation of state organizations where none existed before or the partial or total replacement of existing private ones with their public equivalents. Others were of a contractual nature, such as those related to price formation. These innovations altered some of the coordinating mechanisms that held the system together, modifying some of the relationships between its participants. Most changes of both types were enacted in late 1979 or within the first year of the new government's administration.[41]

Among the institutional changes was the creation of about twenty state-owned corporations. A number of these were granted exclusivity rights in the exercise of their functions:

1. The Nicaraguan Enterprise of Farm Inputs (ENIA), as the sole importer of most fertilizers, insecticides, and other agricultural inputs needed for cotton and other crops. The stated purpose for ENIA's centralization of imports was to secure economies of scale and increased bargaining power in international price negotiations, for the ultimate

---

39. See Henry Ruiz, *El papel político del APP.* See also Eduardo Baumeister, *El subsistema del algodón en Nicaragua* (Managua, INIES, 1983).

40. For an overall assessment of these changes, see, for example, Forrest D. Colburn, *Post-Revolutionary Nicaragua: State, Class, and the Dilemmas of Agrarian Policy* (Berkeley: University of California Press, 1986), pp. 45–61; Baumeister, *El subsistema del algodón;* and Trevor Evans, *El algodón: Un cultivo en debate* (Managua: CRIES, 1987).

41. The nationalization of the banking system was decreed in August 1979; state-owned trading companies, such as ENAL and ENIA, were established in September 1979; state-owned cotton farms and most other state institutions were organized in late 1979 and in 1980; the distribution of cooking oil, a cotton by-product, became state-owned in 1983.

benefit of producers.[42]

2. The National Enterprise of Farm Products (PROAGRO), as the primary domestic distributor of ENIA's imports and, at least initially, the sole supplier of those imports to state farms. The stated purpose of its existence was to ensure the availability of inputs at low prices for the benefit of producers.[43]

3. The Financial Corporation of Nicaragua (CORFIN), which, following the nationalization of the banks, headed all lending institutions dealing with the cotton sector and could ensure credit policy was consistent with food policy objectives.[44]

4. The Nicaraguan Cotton Enterprise (ENAL), as the sole purchaser and exporter of cotton, cotton cake, and linter, was meant to benefit cotton producers by charging them lower export commissions than those private traders had demanded in the past and by obtaining higher export prices through its centralized sales organization.[45]

5. AGROMEC and ETSA, as state-owned companies supplying cotton-farming machinery and plowing and aerial fumigation services.[46]

6. Several large state cotton farms, some of them being the largest cotton plantations in the country, as production entities responsible for generating profit and foreign exchange. These state farms were given most of the expropriated land that had been traditionally used for cotton production.[47]

7. Ten state-owned cotton gins, the result of the expropriation of as many of the 26 private gins operating in 1980.[48]

8. TEXNICSA, a textile plant for cotton spinning and weaving formerly owned by the Somoza family and greatly expanded after it was expropriated by the government.[49]

9. A network of wholesale and retail outlets selling cooking oil and other

---

42. Ministry of Foreign Trade, *Export Directory 1981–1982* (Managua: MICE, 1982); MIDINRA, *Lista de empresas nacionales y de servicios adscritas a las direcciones centrales de MIDINRA y de sus principales actividades económicas al 30 Junio 1983* (Managua: MIDINRA, 1983); Baumeister, *El subsistema del algodón*, pp. 85–86; and DGATM, *Propuesta de fusión ENIA-PROAGRO* (Managua: MIDINRA, 1984).

43. Ministry of Foreign Trade, *Export Directory 1981–1982*; MIDINRA, *Lista de empresas*; Baumeister, *El subsistema del algodón*, pp. 85–86; and DGATM, *Propuesta de fusión ENIA-PROAGRO*.

44. For a description of the structure of the Nicaraguan financial and banking system after 1979, see International Bank for Reconstruction and Development, *Second Agricultural Credit Project, Nicaragua* (Washington, D.C.: World Bank, 1981), pp. 10–18.

45. Empresa Nicaragüense de Algodón, *Memoria, 1979/1982* (Managua: ENAL, 1982); and Ministry of Foreign Trade, *Export Directory 1981–1982*.

46. CIERA/PAN/CIDA, *Informe final*, vol. 3, pp. 51–58; MIDINRA, *Lista de empresas*; and Baumeister, *El subsistema del algodón*, pp. 80, 86.

47. MIDINRA, *Lista de empresas*.

48. Ministry of Industry, *Estudio de la industria de aceites vegetales en Nicaragua* (Managua CIERA/MIND, 1983), p. 159.

49. Ibid., p. 22; and ENAL, *Memoria, 1979/1982* (Managua: MICE, 1982).

cotton by-products, set up after the distribution channels had been nationalized.[50]

Among the contractual changes was the adoption of a pool-pricing system for growers, to be administered by ENAL, the state cotton-trading company. The system involved the setting of a support price in the local currency, the cordoba, every year before the planting season, thereby shielding growers somewhat from international price fluctuations.[51] The pool-pricing system also ensured that all producers would receive the same base price for cotton of comparable quality, eliminating price discrimination among planters as practiced in the past by private traders. The foreign currency obtained from cotton exports was not retained within the cotton commodity system. Instead, it was transferred by ENAL to the Central Bank.[52] Other changes included the setting of lower rates and other controls on cotton land rentals; although rates had reached two thousand cordobas per manzana before the revolution, in 1980 the maximum rate was set at three hundred cordobas per manzana.[53] The 1981 Nicaraguan Agrarian Reform Law was enacted; under certain conditions, it rendered private farms subject to government expropriation.[54] Finally, Sandinista Agricultural Cooperatives (CAS) grouped small cotton producers,[55] and the rural labor union ATC coordinated cotton workers. [56]

High on the long list of benefits that these changes were intended to ensure were the protection of the peasants and rural workers, the profitability of state and private producers, the establishment of state-owned agribusiness enterprises as the overall coordinators of the cotton economy, and the availability of increasing amounts of foreign exchange for use in the expansion of the domestic food system. Peasants and rural workers were to enjoy improved working conditions, greater access to land, and permanent employment

---

50. Ministry of Industry, *Industria de aceites vegetales*, p. 172.
51. A detailed analysis of this pricing system is made by Carlos G. Sequeira, in "State and Private Marketing Arrangements in the Agricultural Export Industries: The Case of Nicaragua's Coffee and Cotton" (Ph.D. diss., Harvard Business School, 1981).
52. Interview with ENAL administrators, January 1987.
53. Decree nos. 230 and 263 of January 7 and February 2, 1980. For cotton land rental rates before 1979, see Instituto Histórico Centroamericano, *ENVIO*, p. 4c; and Williams, *Export Agriculture*, p. 212.
54. Nicaraguan Agrarian Reform Law, decree 782, 1981.
55. MIDINRA, *Tipos de agentes sociales involucrados en la producción de algodón* (Managua: MIDINRA, 1985).
56. Sequeira, "State and Private Marketing Arrangements," p. 149.

through the presence of the union, the process of land reform, the creation of cooperatives for small producers, and favorable new policies of credit and farm inputs adopted by state banks and state distributors. The profitability and security of cotton producers, who often rented some acreage, was to be enhanced through lower land rental rates, guaranteed base prices for cotton, the elimination of price discrimination among producers, lower input costs and higher export prices through state centralization of the purchasing and export functions, and fewer seasonal labor shortages by means of the government-supported mechanization of cotton harvesting.[57]

Finally, the presence of the newly created state farms in the traditionally all-private cotton production sector was to provide the APP with additional sources of profit accumulation, while other state agencies would gain control over the foreign exchange generated by the cotton system, thereby ensuring that both profit and foreign exchange would be used in a manner consistent with food policy goals and the welfare of the needier segments of society. On the whole, the intended simultaneity of these benefits was meant to provide the cotton system with precisely the kind of balanced growth which had been lacking in the past and which was now called for by the new food policy.

Financial Results

The potential conflict between food policy and export agriculture having been resolved, state cotton farms quickly proceeded to expand the acreage devoted to cotton.[58] Their motivation, at first, was a desire to become a growing and increasingly influential state participant in what had been, until 1979, a major center of private profit accumulation. State cotton farms, whose land came partly from that expropriated from large cotton landowners of the northwestern plains and was some of the country's most fertile and best suited for

---

57. For a comparative analysis of cotton export price spreads before and after the introduction of ENAL, see Sequeira, in "State and Private Marketing Arrangements." See also n. 51. For the conflict surrounding seasonal labor, see Laura Enríquez, "Social Transformation in Latin America: Tensions Between Agro-Export Production and Agrarian Reform in Revolutionary Nicaragua" (Ph.D. diss., University of California, Santa Cruz, 1985). For the government's agricultural technology policy and the trend toward increasing mechanization of farm activities, see CIERA/PAN/CIDA, *Informe final*, vol. 3, pp. 51–58.

58. Data sources for this section consist of interviews and the financial reports of state-owned enterprises analyzed during field research carried out in December 1984 and August 1986.

cotton production, rapidly reached prerevolutionary output levels.

When financial records showed that, in spite of their good yields, state cotton farms had incurred losses during their first two years of operation (1980–82), they nevertheless continued to expand cotton acreage, but now their motivation had shifted. The predominant rationale became the need to generate foreign exchange, which was becoming increasingly scarce because of a growing current account deficit (see Table 9), and, by extension, the desire to demonstrate early in the life of the APP that state-owned enterprises could indeed be a reliable vehicle for implementing government policies, even if that implied a financial cost (which, in 1981, was considered transitory).[59]

As cotton expansion continued, however, state farms eventually generated some of the largest financial losses reported by APP enterprises. By 1985, state banks were no longer willing to bear the burden of these large losses. It had become clear that the losses were likely to preclude the economic survivability of state farms as cotton producers; only then was the policy of de facto cotton expansion at any price challenged. State managers started pressing policymakers to permit diversification alternatives away from cotton and into relatively more profitable, nontraded domestic crops, a change they had favored for some time but that only now became compelling enough to attract serious government consideration.

The case of a large state-owned cotton farm located in the main cotton region of Chinandega, henceforth called State Cotton Farm (SCF), illustrates this evolution. SCF started its operations in December 1979. During the preceding months, the Nicaraguan Institute of Agrarian Reform (INRA) had initiated a restructuring of the land formerly owned by Somoza and his associates.[60] As part of the reorganization, INRA created SCF, a state-owned enterprise founded for the purpose of administering approximately fifty confiscated farms representing about 20,000 manzanas (34,000 acres) situated in the best cotton-planting area of Nicaragua. Although most of that acreage had been devoted to cotton in the past, only about 900 manzanas had been planted with cotton in the summer of 1979, a time of broad social disruptions that had preceded the overthrow of

---

59. The 1981 National Economic Plan, for example, states that one of the plan's fundamental objectives for 1981 is the beginning of the process of accumulation (Ministry of Planning, *Programa económico, 1981*, p. 17).

60. For an analysis of the organization of the state agricultural enterprises between 1979 and 1983, see Krüger and Austin, *Control of Agricultural State-owned Enterprises*.

the Somoza regime. In 1979, Nicaragua's total cotton acreage had been one-third of the area planted in 1978, amounting to 64,000 manzanas compared with the average of 260,000 manzanas that had been planted each year between 1974 and 1978 (Figure 3). On the farms assigned to SCF, the cut in cotton planting had been even more drastic, affecting over 95 percent of the acreage, probably because several of their owners had fled the country in anticipation of Somoza's fall, abandoning their estates before the planting season and further disrupting the farms' activities.

Once SCF took charge of these farms, it grouped them into two large cotton "complexes" and began to concentrate its efforts on regaining past production levels. In the following crop year 1980-81, the first under SCF administration, cotton acreage increased fivefold, to 4,480 manzanas. At harvest time the company's average cotton yields came to 14.99 hundredweight/manzana (cwt/mz), equal to the national record of 1972 and 23 percent higher than the national average for 1980-81. The results could in part be attributed to the fact that in 1980, SCF had planted only the very best land at its disposal.

In the following crop year, 1981–82, SCF doubled its cotton acreage compared with the preceding agricultural cycle, to 9,422 manzanas. Average yields came to 12.25 cwt/mz, or 17 percent higher than the national average for that year, and compared well with the national average of 10.94 cwt/mz for the decade of the 1970s. "Although production results were quite acceptable," one SCF administrator explained,

> financial results were negative. In the crop year 1981–82, we lost almost thirty-two million cordobas, an amount that exceeded our company's net worth and that approximated one-half of our total assets. Technically, in 1982 we were already bankrupt. This was not due to a lack of control over our expenditures, since our costs were comparable to those of the private sector. What was happening, instead, was that prices to cotton producers were so low that even with good yields and acceptable costs per manzana, we incurred losses. This policy of low prices was against the interest of the country, since we were then getting better [international] prices than in the late 1970s. At a time when the cotton producer ought to have been given incentives, we were instead discouraged by a policy of low farm-gate prices.[61]

---

61. Interview, August 1985.

72   Hungry Dreams

Figure 3. The structure of the Nicaraguan cotton commodity system: estimated state share of output, 1983

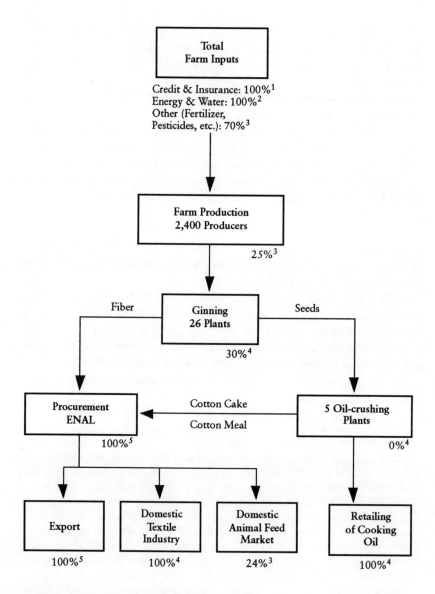

Sources: [1]BCN; [2]INE; [3]MIDINRA; [4]MIND; [5]MICE.

Actually, during the period 1980–82 the government did try to improve prices of export crops. This effort had been motivated in part by the negative financial results of state-owned export enterprises in the crop year 1980–81, the first entirely under the new state administration. The government modified price formation mechanisms and introduced a first package of incentives to export farmers in time for the 1982 cotton crop. "The problem with the incentive program," explained a manager of the SFN involved in its administration, "was that it was insufficient, [it] had been only partially implemented, and in several cases it was not implemented at all. Only some of the dollars needed to cover these incentives had been set aside by the government, and when producers came to claim their incentives, many were turned down because the foreign exchange that we needed to cover their demands had already been spent for other purposes."[62] As a result, the government again modified the system, eliminating the portion that involved foreign exchange disbursements. SCF, however, did not get any of the initial or the modified benefits, and several of its administrators began to base their financial projections on the assumption that these incentives did not exist.

In the crop year 1983–84, SCF reached its peak cotton acreage, with 17,984 manzanas planted, which represented an increase of 35 percent over the previous year and implied that over 90 percent of the land at its disposal had been planted with cotton. By some SCF estimates, this proportion was higher than that reached on the same land prior to the revolution. Yields, however, were 9.68 cwt/mz, or 14 percent below the national average for that year. Negative financial results reached a record level, with losses amounting to more than 115 million cordobas, one of the highest recorded that year by a state-owned enterprise. Only two years after the financial bailout, the company was once again technically bankrupt. But no solution was in sight as long as it continued to grow cotton. "The state does not seem to understand that we are not getting the benefits that they think they are giving us," explained a manager of SCF.

> On the one hand, the CDDs [export incentives] never reached us. They did reach the private sector, though, at least some of the time. So, the private producers who were getting CDDs were actually getting, because of the incentives, a price for their cotton that was 40 percent or 50 percent above

62. Interview, June 1985.

the price we were getting. But we never got those prices. On the other hand, a substantial part of our farm inputs had to be bought at black market prices, thereby nullifying the subsidy that the overvalued cordoba was supposed to give to the cost of our imported agricultural inputs. The state thinks that I am buying [imported] fertilizers at a cordoba price that is related to the official exchange rate of 10 to 1 vis-à-vis the dollar. Instead, the dollar costs me four hundred or five hundred cordobas, which is the exchange rate in the black market. How are we supposed to control our costs when the black market rate is fifty times higher than the official exchange rate? Most of our farm inputs are imported. If I don't pay for them at a cordoba price that reflects that black market exchange rate, frequently I get no spare parts, no fertilizers, and no pesticides. Obviously, I cannot stop production. I have to accept those prices. The net result is that our costs are soaring, while our sales price is kept down when you compare it with the dollars that ENAL gets on the international market for our cotton. Almost everybody is in that situation. But if a private producer doesn't get export incentives or has low yields due to the inferior quality of his land, he can still say, Well, if I have to lose money with cotton, I might as well switch to other things. And that is exactly what many private cotton producers are doing. This is why many former cotton planters don't want to produce exports any longer. But we, as state farms belonging to the APP, have no choice. We have to produce cotton because this is what we are told to do. The plan says plant so many manzanas of cotton. And that is what we do. And so we are forced to lose money. We know it beforehand. But we must follow the directives that we are given. The country needs foreign exchange. But it would be much better for us to produce sorghum. For us, it would mean lower costs and better prices.[63]

Partly as a result of state enterprise losses, the government again changed the pricing mechanism for the crop year 1984–85 and introduced a new package of export incentives for farm producers. But SCF benefited only marginally. Also, in 1984–85 the company, for the first time since its founding in 1979, reduced the area planted with cotton to 15,726 manzanas, or 13 percent less than the previous year, primarily because the state lending institutions were unwilling to finance ever-growing cotton losses. The acreage that SCF could plant de facto was limited by the amount of credit the state bank was willing to give. "Tensions with the bank had already grown in 1982 and 1983 because our sales revenues were unable to cover the loans that the bank was giving us," said one SCF administrator. "But after the financial results of 1984, tensions with the banks became much worse. Actually, our relationship with the banks had

---

63. Interview, December 1984.

deteriorated so much that we had to spend most of our time negotiating new financing instead of managing the farm's operations. Each year since 1981 we were spending less time supervising the work in the field. The constant pressure of the banks took a lot of time and energy from us, and in the end our production suffered."[64] Yields in 1984–85 were no better than the national average, or 9.14 cwt/mz." Losses amounted to about 112 million cordobas and were again among the worst reported by state agribusiness enterprises.

Other state-owned cotton farms reported similar production and financial trends. The area planted with cotton by these farms increased from nineteen hundred manzanas in 1979–80, or 3 percent of the country's total, to more than forty-seven hundred manzanas in 1983–84, or 28 percent of the total. Thereafter, it began decreasing. The cotton output of the private sector, meanwhile, had remained at about half the prerevolutionary level and therefore had clearly not followed the growth path of the state sector. Since the overall production capacity of private cotton growers was significantly larger than that of the state cotton farms, the country's total output during the period 1980–85 was pushed back to the levels of twenty years earlier. This output fall was, in turn, responsible for the country's single largest amount of foregone export earnings since 1980.

The financial losses of state cotton farms appeared all the more disturbing to policy analysts when viewed in combination with the disruptions caused by the private sector's decrease in cotton output and its highly visible ripple effects on employment and on the supply of foreign exchange, cooking oil, and animal feed.[65] Financial difficulties had reached the point of preventing state cotton farms—and state agroexport enterprises in general—from expanding or even maintaining their existing export production levels, an outcome that only a few years earlier had not seemed a distant possibility. It was now necessary to reassess the farms' function as generators of foreign exchange and capital accumulation for the food system, as well as their potential role as substitutes of last resort vis-à-vis what was sometimes perceived as a recalcitrant private export sector.[66] "Before any meaningful evaluation could be made," explained the man-

64. Interview, August 1985.
65. See COSEP, "El algodón, motor histórico de la economía nicaragüense en los últimos 25 años," in *Memorandum de la Presidencia*, no. 9 (Managua: COSEP, April 1986), p. 4.
66. This evaluation, together with other factors that will be mentioned in Chapter 7, eventually led to a new policy that de-emphasized the role of state-owned enterprises in

ager of a state enterprise, "we had to understand what brought us to that situation in the first place. If the government's intention was to give us financial incentives, why is it that we got such low farm prices?"[67]

Uncoupling Farm Prices from the Official Exchange Rate

As noted previously in this chapter, one innovation introduced in the cotton commodity system after the revolution was a pool-pricing mechanism, whose stated purpose was to improve overall prices to producers.[68] Soon after the system was implemented, however, it became apparent that its effectiveness rested on the assumption that the official exchange rate was capable of providing sufficient incentives to motivate cotton growers to produce. In 1981, the government found that this assumption was not so; in order to ensure the profitability of cotton production, it therefore created a new pricing mechanism during the crop year 1981–82 that would not be dependent on changes in the official exchange rate. Henceforth, the pricing mechanism was uncoupled from the official exchange rate. The international price of cotton ceased to be the reference point of farm-gate prices; it was replaced by a once-a-year cost-plus pricing system in cordobas. The new system was unable to provide sufficient farm incentives, however, and was modified several times as policymakers continued to search for a satisfactory mechanism.[69]

Finally, in 1985 the government took a different course of action and devalued the cordoba by 180 percent. In adopting this new approach, the Nicaraguan administration recognized that in recent years an overvalued exchange rate had indeed inhibited production

---

farm production. As a result, after 1985 the percentage of land under control of the APP, according to MIDINRA data, began to decrease.

67. Interview, September 1987.

68. Sources for this section consist of interviews and internal data from state-owned enterprises analyzed during field research carried out between April 1985 and October 1987.

69. See MIDINRA, *Propuesta de política de precios para los productos de agroexportación, ciclo 1982–1983* (Managua: MIDINRA, 1982); MIDINRA, *Precio al productor para el ciclo agrícola 1986/87* (Managua: MIDINRA, 1986); MIDINRA, *Programa nacional algodonero, ciclo 83/83 hasta 85/86* (Managua: MIDINRA, 1982); BCN, *Plan de fomento a las exportaciones*; BCN, *Propuesta de política cambiaria, ciclo 84/85* (Managua: BCN, 1984); and Robert Pizarro, "The New Economic Policy: A Necessary Readjust-ment," in *Revolutionary Nicaragua*, ed. Spalding, pp. 217–232.

and that farm-gate prices had frequently been lower than production costs.[70] At the same time, the government acknowledged the existence of a situation of "chaos in the pricing system, which introduces irrationality in the allocation of products, discourages production, and promotes speculation and smuggling . . . and a parallel [black market] economy which increases at the expense of the formal economy."[71] Accordingly, the government developed an estimate of the shadow price of the cordoba in an attempt to assess the impact that price distortions had been having at the micro level. Managers of state-owned agribusiness enterprises saw in the new government posture a partial vindication of their performance. Many of them felt that in order to prevent a repetition of this experience, however, it was important to clarify why price policies designed for their benefit had had the opposite effect and why it had taken five years to ascertain this problem. A closer look at the pricing mechanisms will address these two questions.

The Cotton Pool-Pricing System, 1979–1981

The adoption of a pool-pricing system implied that each cotton grower was to receive the same base price. This price was to be the cordoba equivalent of the average price in dollars at which Nicaraguan cotton had been sold on the international market. For the system to work, producers had to be paid in two stages.

In the first stage, growers were granted a gross support price in cordobas set every year before the planting season. They received the support price, less a number of deductions, as an initial payment at the moment they delivered cotton to ENAL, the sole purchaser and exporter of Nicaraguan cotton. The deductions were: a price reduction by type of cotton planted (ordinarily, Nicaraguan cotton growers selected from a range of fourteen major cottonseed types); ginning costs; transportation and docking charges; ship-loading fees; cotton classification charges; agricultural tax; and other deductions.

In the second stage, one or more additional payments in cordobas were made, after the completion of the harvest and shipping season, to make the final price in cordobas received by each grower equivalent to the average international price in dollars obtained abroad by ENAL, *after* such international price had been converted into cor-

---

70. SPP, *Plan económico 1985* (Managua: SPP, 1985), pp. 7–8.
71. Ibid., pp. 1–3.

dobas at the official exchange rate. These payments were further subjected to a second set of deductions: ENAL's interest charges on the net support price disbursed, the latter being handled as an advance payment on exports that had yet to be paid by foreign buyers; insurance and warehousing costs between the time of delivery to ENAL and the moment cotton was loaded on a ship; export taxes; and ENAL's commissions.

To cite just one example, in the crop year 1980–81, the gross support price of cotton set in the spring of 1980 prior to planting was 680 cordobas per hundredweight. The final net farm-gate price received the following year at the end of the harvest season, after all additional payments and various deductions had been made, was 613.37 cordobas. The cordoba price f.o.b. the port of Corinto used in government calculations was 789.88 cordobas and represented the gross support price plus all additional payments. The weighted average international sales price actually obtained by ENAL for that crop year was $75.48. At the official exchange rate of 10 cordobas for one dollar, that sales price was equivalent to 754.80 cordobas, a number approximating the final gross "f.o.b." price of 789.88. In other words, the goal of providing to each cotton producer a base price that was the cordoba equivalent of the international price, net of f.o.b. charges, had been achieved within reasonable approximation. Actually, if one were to accept somewhat different calculations, it was possible to argue that the system had functioned with great accuracy.[72]

For all practical purposes, however, that mechanism did not last for more than one year. Although the cotton pool-pricing system had been in effect since the creation of ENAL in September 1979, it was not until early 1981, when the first cotton crop cycle entirely under the new pricing mechanism was completed, that all its features became operative and a government assessment of the system could be made.[73]

According to an SFN official, the evaluation showed that for cotton pool-pricing to be effective, one critical condition had to be met:

---

72. If one were to deduct from the f.o.b. price the agricultural and export taxes, the interest charges, and ENAL's commissions, which arguably do not belong to this price, then the estimated f.o.b. price would be reduced to 757.35 cordobas, that is, almost exactly the export price of 754.80 as converted by the official exchange rate.

73. During the crop year 1979–80, no support price prior to planting was announced, because the new revolutionary government came to power on July 19, 1979—after the

That condition was that the official exchange rate be at a level capable of providing sufficient profitability to producers. That we had to make such a discovery should not come as a surprise if one thinks for a moment about Nicaragua's past record in this respect. After all, the year 1979 had been preceded by two decades where our currency was stable and freely convertible to the dollar, where we had no major trade deficits, and where the inflation rate was around 6 percent and run parallel to that of our major trading partners. When exchange rate distortions surfaced in the early 1980s, we were caught somewhat off guard. But during the crop year 1981–82, it became apparent that the official exchange rate was not providing sufficient profitability to export producers. Something needed to be done about it. Since the government was unwilling to devalue the cordoba, all we could try to do was to design a new system that could provide export incentives without touching the official exchange rate. In other words, for the sake of motivating producers, we uncoupled the pricing system from the official exchange rate."[74]

In fact, in the spring of 1981 the gross support price relating to the 1982 harvest was increased by 23 percent, although the international price of cotton had actually decreased between 1981 and 1982. The support price was increased because the costs in cordobas to producers were increasing. By 1985, domestic inflation was 330 percent according to government estimates; in reality, it was probably higher. In theory, many products could be bought at regulated prices; in practice, they were not available at such prices. The only way to get them was to buy them on the black market, where prices were many times the official price. When, at harvest time, Nicaraguan cotton averaged $64.47 in the international market, ENAL discovered that, the exchange rate being what it was, it could not even cover the support price of 840 cordobas, not to mention making additional payments. It was then that the government realized the system was not going to work.[75]

The "Cost-Plus" System, 1981–1984

From that point on, the burden of making export agriculture attractive was removed to a large extent from the arena of foreign exchange policy and became primarily a domestic price policy re-

---

planting season had already started. The cotton-planting season usually begins in late June and ends in July, with some cotton being planted as late as early August.
74. Interview, June 1987.
75. Interview with ENAL administrative staff, January 1987.

sponsibility. The government designed a cost-plus system in domestic currency, on top of which it added some export incentives in dollars on an ad hoc basis. To accomplish this task, the government created a system of multiple exchange rates, which obviously amounted to an implicit devaluation of the cordoba. But the whole mechanism was not designed to monitor differential inflation, something that a more explicit foreign exchange policy presumably would have done.[76] In 1981 the government thought that Nicaragua had one of the lowest inflation rates in Central America.[77] It later learned that in 1980 the rate of inflation had already exceeded 30 percent and in 1985 was well above 300 percent. Yet the pricing mechanism was a complex cordoba-oriented process based on a once-a-year estimate of average production costs timed around the crop cycle rather than around changes in domestic price levels. Such a mechanism proved far too slow in adjusting farm prices to domestic inflation. By the time the harvest season arrived, the basic cotton price in cordobas was obsolete, and the incentives could not make up the difference. To make matters worse, those incentives, insufficient as they were, sometimes were not even delivered.

In any event, because the pricing system had lost its connection with the official exchange rate, government decision makers lost the habit of inquiring at what dollar price cotton was being sold. Had they raised that question, the overvaluation of the dollar would have been unmasked. Instead, they were working with a price mechanism that had been turned upside down. The question being asked was, Based on our once-a-year production cost estimates in cordobas, how much does the ad hoc (multiple) exchange rate for crop x, or crop y, have to be in order to make its production profitable? The

---

76. The Central Bank of Nicaragua did have calculations of the relationship between the consumer price indexes of Nicaragua and the United States. These calculations showed that between 1971 and 1978, the relationship was constant and almost equal to zero, which indicated nearly equal price variations between the two indexes. The ratio began increasing in 1979, as Nicaraguan domestic prices rose faster than did U.S. prices. It reached a level of about 30 in 1984, jumped to about 80 in 1985, and almost reached the level of 500 in 1986. These data, however, were not included in the ordinary price formation decision-making process during the period 1980–85 (interview with a manager of the National Financial System, November 1987).

77. The 1981 National Economic Decision Plan stated that "the inflation rate for the year 1980 is one of the lowest in Central America" (Ministry of Planning, *Programa económico*). On the other hand, data from CEPAL indicate that in 1980, Nicaragua had the highest rate of consumer price increases in Central America, equal to 24.8 percent (M. E. Gallardo and J. R. López, *Centroamérica: La crisis en cifras* [San José: IICA/FLASCO, 1986], p. 132). Likewise, estimates of the Central Bank of Nicaragua show that in 1980 the country's inflation rate was 35.3% (see Table 9).

government would then work out a system of incentives that would produce that exchange rate. The problem, however, was that the cordoba, not the dollar, was the starting point. And the cordoba proved to be a moving target. By the time the government had "shot" a cordoba price for the crop year, the target was elsewhere. By harvest time, costs were once again higher than revenues. Because the price mechanism was causing problems, the government kept changing its nature, not its speed of adjustment. The government was blaming state enterprises for having little control over costs, but it did not take domestic inflation seriously. In hindsight, it is clear that the cordoba was appreciating faster than the pricing system could selectively devalue it.[78]

The crop year 1981–82 will again illustrate the point. As was shown in the previous section, had the pool-pricing system been left unchanged, at the end of the 1982 cotton harvest the final gross f.o.b. price would have been 840 cordobas, since ENAL was not in a position to make payments in addition to the initial support price. This situation existed because in early 1982, ENAL had sold its cotton at $64.47, that is, at 644.70 cordobas according to the official exchange rate. Under the circumstances, ENAL was supposed to honor the support price of 840 cordobas that had been established before the 1981 planting season. But honoring this price would have involved the disbursement of a subsidy of 195.30 cordobas per hundredweight, the difference between the support price and the converted export price. Because ENAL had no funds of its own (its export receipts were transferred to the Central Bank and its operating budget was financed only by commissions) and because the support price of 840 cordobas, which resulted in a farm-gate price of 709.56 cordobas, was considered too low anyway, the government decided to change the system.

The announcement was made in February 1982, far too late to affect the acreage planted in 1981–82 but early enough to solve ENAL's dilemma and answer in part the growing financial difficulties of producers, which were likely to affect the next planting season. A MIDINRA internal position paper prepared in August 1982 summarized the situation in a section devoted to cotton price policy.

> The policy of revising the costs of production [of cotton growers] and updating the prices [of cotton] aimed at guaranteeing a level of profitabili-

---

78. Interview with a MIDINRA manager, August 1986.

ty that would motivate the producer.... The decision to increase the prices almost at the end of the agricultural cycle [February 1982] and therefore without the possibility of influencing it, is based on the reasoning that the financial results [of producers] for that cycle 1981–82 would influence in a decisive way the amount planted in 1982–83.... If we consider the historical behavior of cotton growers and their current level of indebtedness, a promise of profits in the next cycle [1982–83] while they lose on this one [1981–82] is not enough and would result in a reduction in cotton production. The dependence of producers on credit to finance their working capital, and their level of indebtedness, would force many producers to look for alternative ways of making a profit, or even to exit altogether from production.[79]

The new system had two major components. First, the cost-plus method determined before the planting season the domestic support price of cotton for the forthcoming crop year. Second, the Certificates of Exchange Availability (*Certificados de Disponibilidad de Divisas*, or CDDs) were applied at the end of the harvest and were introduced just in time for their first application to the 1981–82 crop.

The "Cost-Plus" Pricing System. A commission composed of members of the Ministry of Agriculture and Land Reform (MIDINRA), the Ministry of Foreign Trade (MICE), the Ministry of Planning (MIPLAN), and the Central Bank of Nicaragua (BCN) administered this system. In order to set the price of cotton before the planting season, the commission first had to estimate the average cost of preharvest operations per manzana. This average cost was based on a sample of cotton farms whose extension exceeded seventy-five manzanas, and for all practical purposes, it was treated as a fixed cost, since it was incurred independently of the cotton yield that would be obtained at harvest time. For the crop year 1982–83, this cost was estimated at 8,217.10 cordobas. The second step was to estimate the average yield per manzana, before and after ginning. In the case of the crop year 1982–83, the estimate was 37 cwt/mz and 12.39 cwt/mz, respectively. The third step was to estimate harvesting and ginning costs per hundredweight, which in this case were 62.50 and 168.72 cordobas, respectively.

Once the commission made those estimates, it was possible to establish the projected cost of one hundredweight of ginned cotton. Thereafter, the government was in a position to decide and eventually announce, before the planting season, the postharvest price ENAL

---

79. MIDINRA, *Propuesta de política de precios*, pp. 14–15.

would pay producers for their cotton. Essentially, the domestic price of cotton was now a function of its expected cost in cordobas and of the profit margin that the government would grant. It had lost its relationship with the international price of cotton. In the particular case of the 1982–83 cotton crop, this domestic price was set at one thousand cordobas per hundredweight of ginned cotton. The government set this price with the intent of giving planters a profit, as a percentage of their total costs, of 6 percent, *provided* that (1) those costs were equal to the national average as estimated by the government prior to planting and (2) their yield before ginning was 37 cwt/mz. The government also estimated that a planter whose costs were equal to the estimated national average would reach the break-even point with a yield of 34.32 cwt/mz of pre-ginned cotton.

This system was introduced in February 1982; therefore it could not address the particular problems encountered with the price of the 1981–82 crop.[80] As we shall see in the following section, those problems were meant to be solved by the CDDs, which constituted the other component of the new pricing system.

Certificates of Exchange Availability. "The easiest way to show what the CDDs were for cotton and how they worked," explained an SFN government official, "is to go through an actual example."

> Let us assume that we are in the crop year 1982–83 and that we have the case of a hypothetical cotton producer, Mr. X. At planting season in June 1982, Mr. X already knows that the price of ginned cotton for that season shall be 1,000 cordobas per hundredweight. Let's assume now that at the end of his harvest, in January 1983, Mr. X has produced 100 hundredweights of ginned cotton and that he has incurred in exactly the same unit costs and yields as those that the government had estimated eight months before, that is, prior to planting, as national averages. His total sales revenues will be 100,000 cordobas. His profit will be 6 percent of the total cost of producing that ginned cotton, or 6,111 cordobas.
>
> At this point let us introduce the "cotton CDD for 1982–83." I call it the "cotton CDD for 1982–83" for two reasons: first, because there is a different type of CDD for each export product, and therefore the CDD that applies to cotton does not apply, for example, to coffee or sugar; and second, because the CDD of the same export product, in our case, cotton, may vary from one crop year to the other. In other words, the cotton CDD

---

80. In February 1982 the support price of cotton for the crop year 1981–82 had already been established at 840 cordobas and had been effective for almost one year. The substitution of a higher one, based on the "cost-plus" method, in and of itself would have only compounded ENAL's problems and its inability to pay cotton growers.

for 1982–83 is not necessarily the same as the cotton CDD for 1983–84. Henceforth, however, we shall call this "cotton CDD for 1982–83" simply a CDD. The CDD defines the amount of profit that can be converted by a cotton grower into dollars. This amount is expressed as a percentage of sales. In our case, the CDD is 61 percent, meaning that all the profit that a grower makes that is not in excess of 61 percent of his sales revenues can be converted into dollars. The conversion rate is the official exchange rate, that is, 10 cordobas to the dollar. Since the profits of Mr. X are equal to the estimated national average of 6 percent of total costs, they amount to 6,111 cordobas. As a result, all his profit can be converted into dollars, since 6,111 represents only 6 percent of his sales, and he will get $611.10 for it. Actually, he doesn't get the dollars, but a Certificate of Exchange Availability, or CDD, for that amount.

At this point the system starts to complicate. As originally designed, [the system gave] Mr. X . . . two options. One option was to use those CDDs to import anything that he wanted without the need for a special import license and without having to go through cumbersome official channels. In a situation like the one we were in, with scarcity of foreign exchange and shortages of both consumer and capital goods in the domestic market, that possibility was a great incentive.

The other option was to reconvert the CDDs into cordobas, this time, however, at the special exchange rate of 28 cordobas to the dollar. That means that the $611.10 that Mr. X held in CDDs would be reconverted into 17,110.80 cordobas, thereby allowing him to make an "exchange rate differential profit" of 10,999.80 cordobas. If this amount was added to the cost-plus sales revenues of 100,000 cordobas, in the final analysis it meant that Mr. X had been paid 1,109.99 cordobas for each hundredweight of ginned cotton. Since ENAL in early 1983 sold the cotton crop at an average price of $63.45 per hundredweight, the implicit exchange rate, that is, the ratio between the domestic price in cordobas and the international price in dollars, was 19.80, or a de facto selective devaluation of the cordoba of 98 percent. Naturally, if Mr. X had made no profit, either because his costs and/or his yields had been below the government-established average break-even point, then Mr. X was getting no CDDs.

Going back to the crop year 1981–82, the CDDs managed to raise the support price of cotton by 124 cordobas, so that the final gross f.o.b. price of cotton was 964 cordobas. This appeared to have solved the problems of both ENAL and the producers. Since ENAL had sold its cotton that year at $64.47, which converted at the official exchange rate was equivalent to 644.70 cordobas, the government stressed that anything between that amount and the actual domestic price of 964 cordobas was indeed a government subsidy to producers. At that time there was little doubt in our mind that we were doing a good job at accommodating producers' needs.[81]

81. Interview with an officer of the SFN, June 1985.

The Government's Evaluation of the "Cost-Plus" and CDD System. The government's objective in establishing this system was to decrease the current account deficit and restore the profitability of export production, which had been negatively affected by domestic inflation.[82] The BCN reported, however, that "after two years of implementation . . . the system had not succeeded in increasing exports."[83] In particular, the area devoted to cotton, on the whole, had not increased but had remained at approximately half its pre-revolutionary level. Why had the system failed to increase exports or restore profitability?

If we first look at the cost-plus system, we see that the assumptions on which it was based proved critical. For example, the system implied that in 1982–83 the break-even point for a producer with costs equal to the national average was obtained with a pre-ginned yield of 34.42 cwt/mz. That yield was higher than in four of the five crop years covering the period 1974–79. In effect, by setting a high break-even point, the government was excluding beforehand many producers at a time when the policy was to increase cotton production.[84] That problem was compounded by what happened with the CDDs.

First of all, the weighted average multiple exchange rate increased from 13.70 in 1982 to 15.47 in 1983 because of CDDs, while many imports remained at the official exchange rate of 10 cordobas to the dollar. The result was a significant exchange rate loss for the Central Bank. By December 1983 the difference between the prices at which the bank bought and sold dollars under the de facto multiple exchange rate system created by CDDs amounted to 2.6 billion cordobas. This loss contributed to an increase in the level of liquidity of the economy and, consequently, the level of inflation.[85]

As a result, the Central Bank was unwilling to finance the CDDs, which would have meant more losses for the bank, more paper money, and more inflation. Increased inflation would mean that more CDDs would be needed the next time to motivate producers. When the CDD system was introduced, no cash flow or financial projection had been made to ensure that the dollars would be avail-

---

82. MIDINRA, *Propuesta de política de precios;* MIDINRA, *Programa nacional algodonero;* BCN, *Plan de fomento a las exportaciones;* BCN, *Propuesta de política cambiaria.*
83. BCN, *Propuesta de política cambiaria,* p. 8.
84. Interview with a manager of state-owned enterprises, August 1986.
85. BCN, *Propuesta de política cambiaria,* pp. 11–12.

86   *Hungry Dreams*

able to cover the CDDs. So the rules concerning CDDs kept changing as cotton planters came to claim them. First they were told that CDDs were good for importing vehicles and agricultural equipment, later, only for equipment, and finally, not even for equipment. Then MIDINRA volunteered to make its new jeeps available to help producers out of the impasse. Producers started to exchange CDDs for jeeps. Soon, however, the supply of jeeps ran out, and the CDD entirely lost its connection with the dollar and with the ability to import. It could be used only for reconversion into cordobas. As a result, many producers became disenchanted with export incentives and began to look for ways to make money in the domestic market, in place of cotton production.

More important, perhaps, with the introduction of the CDD, the government lost control of foreign exchange policy. Since no one could really predict how many cotton planters were going to produce above the break-even point, policymakers had no way of telling in advance how many CDDs would be issued. It was only eighteen months after the initial cost-plus price had been announced, that is, only after knowing at what weighted average price ENAL had bought and sold its cotton and what had actually happened with the CDDs, that the government could calculate the weighted average of all export operations to find out by how much the cordoba had been devalued. Another problem was that policymakers lost track of the change in the relative prices of export products created by differentiated CDDs, which in turn produced another distortion. The export products that generated the greatest amount of net foreign exchange were not those given higher CDD percentages.[86] Instead of encouraging the export of those products that were more dollar-efficient, the government often did the opposite, such as en-

---

86. By net foreign exchange, or *generación neta de divisas* (GND), of a given product X is meant the difference between the amount of dollars that the export of one unit of product X generates and the amount of dollars that the country spends on imported inputs needed for the production of that unit of product X. A product with a positive net foreign exchange generation gets more dollars in exports than it consumes in imported inputs. To this net amount called GND the government sometimes adds, on a pro rata basis, the amount of dollars earned by the export of the by-products of X and/or the dollars saved by not having to import X and its by-products if X were not produced domestically. In the case of cotton, for example, some calculations of the GND include as a revenue the value of cottonseed, cotton linter, and cotton cake imports, as well as the estimated dollars that Nicaragua saves through cotton production by not having to import cooking oil.

couraging sugar instead of coffee.[87] Ultimately, then, the selective devaluation of the cordoba had been de facto an overvaluation of the cordoba. At this point the government decided to scrap the system and replace it in the 1984–85 crop year with the guaranteed price system, which included the actual delivery of dollars in cash into the hands of producers. Finally, in February 1985, the govern-

---

Because these methods of calculation are not applied by the government in a uniform way, discrepancies are frequent between the results of one GND calculation and another. In part, these inconsistencies explain why, after 1985, a controversy surfaced among government analysts as to which was the real amount of net foreign exchange that cotton generated and whether it was positive, as most analysts had believed until 1985, or negative, as a few analysts began to believe after 1985. From some publications of the private sector federation, COSEP, it could be inferred that the position of private cotton planters was that cotton's net foreign exchange was obviously positive. Several government officials of different state institutions also expressed to me this same opinion. And all the GND calculations made by MIDINRA between 1980 and 1985 showed a positive cotton GND. What clouded the issue, however, were a number of other factors that prevented the settlement of the dispute. One such factor was the fact that all GND calculations estimated the amount of dollars needed to produce one unit of cotton by listing the actual price in cordobas that a given cotton producer had to pay for its imported inputs. This calculation ignored the fact that these prices in cordobas, such as the price of fertilizers and pesticides, included the domestic distributors' gross margin, which does not involve a cost in dollars. Since these gross margins came out to be quite substantial (in 1984, for example, the state-owned farm supplier PROAGRO had a cost of goods sold that amounted to only 17 percent of its sales volume), much of the cordoba price of the imported farm inputs was inappropriately converted into dollars. A calculation that would take this point into account would obviously show a much more positive GND and would have probably removed most of the doubts that still lingered. The fact that major importers and distributors of agricultural inputs, such as ENIA and PROAGRO, had their financial accounting reports lagging by six to eighteen months, however, often made extremely difficult, if not impossible, a timely and accurate estimate of GND before the cotton-planting season, that is, when the question was raised of how much cotton the government should encourage producers to plant. A second element that complicated the issue was the question of what exchange rate had to be used for the conversion of the cordoba prices into dollars. Analysts had at least six different types of exchange rates to choose from. More choices had to be made within a single type of exchange rate. Given the number of assumptions that any such calculation would contain, few cotton GND estimates were alike, an outcome that preserved the uncertainty as to which was the cotton GND and more importantly, which was the appropriate agricultural policy regarding cotton. Part of this debate around cotton is reflected in a series of front-page articles published by *Barricada* on February 2, 3, and 4, 1987. See also, Eduardo Estrada and Bismark Bodan, "El algodón: Rentabilidad o alta producción," in *Boletín socio-económico*, no. 1 (Managua: INIES, January-February 1987), pp. 5–8. Additional elements on this issue can be found in MIDINRA, *La generación neta de divisas del sector agrícola y agroindustrial de exportación* (Managua: MIDINRA, 1983); MIDINRA, *La generación neta de divisas en el sector agropecuario* (Managua: MIDINRA, 1985); COSEP, "El algodón," p. 4; and Trevor Evans, *El algodón: Un cultivo en debate* (Managua: CRIES, 1987).

87. Government calculations showed that coffee's net foreign exchange generation (GND) was much higher than that of sugar (see MIDINRA, *La generación neta de divisas en el sector agropecuario*, p. 4).

ment decided to devalue the official exchange rate vis-à-vis the dollar by 180 percent.[88]

The Impact of an Overvalued Exchange Rate on the State Cotton Farm

Between 1980 and 1985, five different exchange rates were operating simultaneously in Nicaragua. Exchange rates were always expressed in terms of the number of cordobas needed to buy one dollar. At one end of the spectrum was the official exchange rate, which during this period had remained fixed at 10 cordobas to the dollar, until the devaluation of February 1985 brought it to the level of 28 cordobas. At the other extreme was the illegal but widely used black market rate, which competed with the government for the availability of increasingly scarce dollars and came under systematic monitoring by the Central Bank after August 1981. This rate went from 21 cordobas in 1980 to 1,000 cordobas at the end of 1985.

Somewhere between those two values, but much closer to the black market rate than to the official exchange rate, operated a third value, the so-called parallel rate. This value, along with the private exchange houses that honored it, was regulated by the government and tried to capture part of the black market dollars without hoping to replace that market entirely.[89] Foreign exchange operations authorized at the parallel rate included transactions such as the purchase of dollars for the payment of medical supplies, spare parts, and farm equipment acquired abroad by local residents and the sale of dollars and other foreign currencies by visitors and tourists to cover their living expenses while in Nicaragua. This parallel rate went from 21 cordobas in the fall of 1981 to 770 cordobas at the end of 1985, and during the periods that it was allowed to function, it followed the changes in the black market rate relatively closely.

A fourth exchange rate, called the implicit exchange rate, was the relationship between the price in cordobas f.o.b. the port of Corinto of a given export product belonging to a given producer and the price in dollars at which this product was actually sold on the international market by one of the state trading companies responsible

---

88. Interview with an SFN manager, June 1985.
89. BCN, *Establecimiento de un mercado paralelo* (Managua: BCN, undated). The document was written between 1979 and 1983.

for its export. Because of the CDD system, after 1981 there was one implicit exchange rate for each product, which varied annually and by individual producer. In the case of cotton for the hypothetical farmer X, the implicit exchange rate in the crop year 1982–83 was 19.80, or 98 percent above the official exchange rate. In the case of the state farm SCF, it had varied between 13 and 30 cordobas (Table 12, line 9). This implicit exchange rate roughly measured the outcome of the government's efforts to selectively devalue the cordoba in order to restore the profitability of export production.[90] The concept of implicit exchange rate was particularly useful for identifying changes in relative prices among export products. For example, some studies based on this concept showed that between 1979 and 1983, coffee's implicit exchange rate had been so low compared with other products as to be even lower than the official exchange rate of 10 cordobas to the dollar, which implied a possible discrimination against that crop.[91] Other studies showed that the export products most favored by the pricing policy were neither those with the best international price prospects nor those that generated the highest amount of foreign exchange per dollar invested.[92]

A fifth exchange rate, sometimes called the "multiple" exchange rate, was the weighted average for a given product or group of products of all the various implicit exchange rates created by the introduction of the CDDs in February 1982 and the application of a selective system of import and export taxes.[93] According to estimates by the International Monetary Fund (IMF) on the basis of data supplied by the Central Bank, its value went from 15.60 in 1982 to 19.18 in 1984 for cotton and from 13.70 in 1982 to 17.90 in 1984 for export goods as a whole.[94] The government had to use eleven

90. BCN, *Propuesta de política cambiaria.*
91. See, for example, Francisco Mayorga, *Nicaragua: Trayectoria económica, 1980–1984—Algunas apreciaciones* (Managua, 1985), pp. 49–59; Colburn, *Post-Revolutionary Nicaragua,* pp. 62–84; Carlos Molina M., Orlando Morales O., and Oscar Neira C., *Los términos económicos de la alianza obrero-campesina: El caso de los precios relativos* (Managua: CIERA, 1985), p. 10; José Luis Medal, *Políticas de estabilización y ajuste estructural en Nicaragua, 1980–1986* (San José: CSUCA, 1987), pp. 26–35.
92. For example, see MIDINRA, *La política de precios agropecuarios* (Managua: MIDINRA, 1987), p. 12. See also Molina, Morales, and Neira, *Los términos económicos,* p. 10; Medal, *Políticas de estabilización,* pp. 226–35.
93. The IMF describes this exchange as being "the weighted average of the estimated value of transactions; the types of exchange rates for the export of goods constitute mixed types which result from the various regulations for the payment of revenues that come from exports at the official and parallel exchange rates" (IMF, *Nicaragua: Evolución económica reciente* [Washington, D.C.: IMF, 1985], p. 129).
94. IMF, "Tipos de cambio implícitos," in *Nicaragua: Evolución económica reciente,* p. 130.

different multiple exchange rates simply to estimate the amount of dollars spent on imported agricultural inputs needed to grow one manzana of cotton. Since no complete and updated list of these multiple exchange rates was readily available, two government institutions using different sets of multiple exchange rates would reach different conclusions on similar problems, an occurrence that would create considerable uncertainty and a tendency to postpone important decisions.[95]

In 1985 the government introduced a sixth exchange rate for its own internal use. This rate was an estimate of the shadow exchange rate of the cordoba as calculated by several government institutions.[96] The development of these estimates aimed at assessing the direction and to some degree the extent of the price distortions that the overvalued exchange rate was having on economic results.[97] The shadow exchange rate estimated by the National Investment Fund (FNI), the organization established in December 1983 to centralize all foreign borrowing for development purposes, placed the value of the cordoba at 20.23 in 1980 and at 198.00 in 1985. Its value over time, therefore, increasingly distanced itself from the official and the multiple exchange rates, though it remained closer to them than to the government-regulated "parallel" rate or the value in the black market.

Table 12 is a first approximation, based on those shadow exchange rates, of the impact of an overvalued exchange rate on SCF. The first step was to calculate the amount of incremental revenues that SCF would have received had the official exchange rate been equal to the shadow exchange rate. This amount, reported in Table 12, line 13, is the "export tax due to overvalued exchange rate." The second step was to estimate the subsidy on imports that was

---

95. A typical example is the case of cotton. The issue is discussed in n. 86.

96. Given are the basic documents that supported this effort: National Investment Fund, *Factores de conversión para Nicaragua* (Managua: FNI, 1985). IMF, "Indices de tipo de cambio efectivo," in *Nicaragua: Evolución económica reciente*, p. 48a. National Institute of Energy (INE), *Un nuevo enfoque en la determinación del precio de cuenta de la divisa. El caso de Nicaragua* (Managua: INE, 1984). Ministry of Planning, *Cálculo de los factores de conversión a precios de eficiencia económica para Nicaragua* (Managua: MIPLAN, 1983). Interamerican Development Bank, *Los precios sociales en Nicaragua: Dos alternativas de estimación* (Managua: IDB, 1977). IDB, *El precio social de las divisas en Nicaragua* (Managua: IDB, 1976). See also IMF, "Indices of Effective Exchange Rates and Relative Prices," in *Nicaragua: Recent Economic Developments* (Washington, D.C.: IMF, 1986), p. 46a.

97. FNI, *Factores de conversión;* INE, *Determinación del precio;* and Ministry of Planning, *Cálculo de los factores.*

Table 12. State Cotton Farm: adjusted profit (loss), 1981–1985

		Measure	1981–82	1982–83	1983–84	1984–85
1	Area planted[a]	mz	9,422	13,281	17,984	15,726
2	Area harvested[a]	mz	9,422	13,056	16,915	15,145
3	Ginned cotton produced[a]	cwt	115,445	206,640	163,721	138,356
4	Ginned cotton yield (harvested)	3 + 2	12.25	15.83	9.68	9.14
5	Average farm-gate price	C$/cwt	709.56	794.56	1,207.95	1,467.12
6	F.o.b. charges[b]	C$/cwt	160.44	146.31	196.84	359.03
7	F.o.b. price	5 + 6	870.00	940.87	1,404.79	1,826.15
8	Export price[b]	$/cwt	64.58	63.36	73.39	62.37
9	Implicit exchange rate	7 ÷ 8	13.47	14.85	19.14	29.28
10	Official exchange rate[c]	C$ for $1.00	10.00	10.00	10.00	10.00
11	Shadow exchange rate[d, e]	C$ for $1.00	25.12	23.59	120.00	198.00
12	Difference between shadow and official exchange rate (C$)	11 − 10	15.12	13.59	110.00	188.00
13	Export tax due to overvalued exchange rate (C$ 000)	12 × 8 × 3	112,726	177,930	1,321,703	1,622,302
14	Imported inputs per manzana[a,f]	$/mz	254.37	269.29	342.06	484.67
15	Imported inputs per cwt ginned ($/cwt)	14 ÷ 4	20.76	17.01	35.34	53.03
16	Official exchange rate for imported agricultural inputs[c]	C$ for $1.00	10.00	15.00	15.00	20.00
17	Difference between shadow and official import exchange rate (C$)	11 − 16	15.12	8.59	105.00	178.00
18	Imported subsidy due to overvalued exchange rate (C$ 000)	17 × 15 × 3	36,237	30,193	607,519	1,305,989
19	Export tax less import subsidy (C$ 000)	13 − 18	76,489	147,737	714,184	316,313

Table 12. Continued

	Measure	1981–82	1982–83	1983–84	1984–85
20	SCF profit (loss) before adjustment[a]	C$ 000	(31,904)	(17,454)	(115,867)	(112,223)
21	SCF profit (loss) after exchange rate adjustment (C$ 000)	20 + 19	44,585	130,283	598,317	204,090
22	Interest on SCF loans[a]	C$ 000	150	4,819	12,883	25,447
23	Average interest rate[a]	%	19	19	19	22
24	Inflation rate (average of two years)[c]	%	28	28	42	192
25	Adjusted interest rate	23 + 24	47	47	61	214
26	Interest subsidy (C$ 000)	Principal × 24	219	7,016	28,132	218,903
27	SCF profit (loss) after exchange and interest rate adjustments	C$ 000	44,366	123,267	570,185	(14,813)

Sources: [a]SCF; [b]ENAL (Nicaraguan State Cotton Company); [c]BCN; [d]FNI; [e]INE; [f]MIDINRA.
Note: C$ = cordoba; $ = U.S. dollar.

due to the overvalued exchange rate.[98] This amount, reported on line 18, was obtained using government estimates and SCF calculations of the dollar component of each cost of producing cotton.

The third step was to remove from SCF's losses, as reported in its financial statements, the positive difference between the export tax and the import subsidy. The net result is shown in Table 12, line 21, and is the "SCF profit after exchange rate adjustment." If we compare these adjusted profits in line 21 with those originally reported by SCF (line 20), we see that, as a first approximation, SCF shifts from a situation where substantial losses are reported every year to a situation where just the opposite occurs and the company reports substantial profits each year. "We knew that we were making a profit," explained an SCF manager, "but our problem was that we did not have the skills to develop shadow prices on our own. All we had as a reference was the parallel rate and the black market rate. Of course, everybody agreed that those rates were too high, even the parallel one which had the government's blessing, because they were in part the result of the activity of speculators. But everybody also agreed that the official and the multiple exchange rates were too low. But the basic point was that anywhere in between those rates, we would have been profitable."[99]

A more realistic estimation of SCF financial results would entail removing the effect of subsidized interest rates on costs. Chapter 6, which covers credit policy, notes that interest rates on agricultural loans between 1980 and 1985 usually ranged between 12 percent and 20 percent. As the inflation rate during the same period was higher, most interest rates were negative. In the case of SCF, a rough approximation of this implicit subsidy is given in Table 12, line 26.[100] Once this subsidy is removed from the profits resulting from the shadow exchange rate adjustment, we see that SCF would continue to remain profitable except in 1985, when the company would show a small loss (Table 12, line 27).

98. Since the cordoba was overvalued, imports were implicitly subsidized. A significant proportion of imported inputs, however, were sold in the domestic market at a cordoba price that reflected the black market exchange rate. These prices were partly due to the development of a black market for agricultural inputs and a corresponding shortage of inputs at official prices. See Pizarro, "New Economic Policy," pp. 221–223.
99. Interview, December 1986.
100. In this first approximation, the interest subsidy for the crop year 1984–85 is considerably overestimated, because in 1985 most of the increases in the inflation rate occurred in the second half of that year and therefore are not relevant for the time period covered by Table 12. That fact may explain in part why in the last adjustment of SCF's financial results, the company shows a net loss for 1984–85 (Table 12, line 27).

94   Hungry Dreams

In 1986 and 1987 the domestic inflation rate exceeded 1,000 percent, while interest rates remained between 25 and 45 percent. This discrepancy effectively subsidized highly indebted state enterprises, far overcoming the tax represented by a still overvalued exchange rate. Partly as a result of that circumstance, after 1985 many state agribusiness enterprises began reporting huge profits. "Just as the previous losses between 1979 and 1985 had often been paper losses, and perhaps state enterprise managers were unjustly blamed for them," explains a MIDINRA manager, "so the huge profits that many state companies began reporting in 1986 and 1987 were artificial profits too. The former led to too much pessimism, the latter to a short-lived euphoria. Both proved harmful to the proper administration of state enterprises' assets."[101]

A Survival Strategy: Switching from Exports to Domestic Production

The devaluation of the cordoba, announced on February 8, 1985, signaled a new direction in the government's economic strategy. Given the government's past reluctance to adopt this measure, understandable doubts as to its appropriateness lingered among many public officials.[102] A few weeks after those measures were announced, an economist of the Ministry of Foreign Trade, Roberto Pizarro, wrote:

> The task of explaining this decision is critical in order to allow for the development of a national commitment to the restructuring of the country's economy. . . . It is probable that the adoption of a readjustment package in 1981, when some of the recent problems were just appearing, would have partly relieved the economic tensions that have accumulated. . . . It would be wrong to blame all the economic difficulties on factors outside the control of the revolutionary government. . . . The need to raise foreign exchange through an aggressive policy of export production and the concomitant need to effect savings by adopting a policy of import rationalization were not met. . . . The maintenance of the overvalued cordoba in rela-

---

101. Interview, October 1987.
102. The government's reluctance to proceed with an outright devaluation of the cordoba between 1980 and 1984 was partly motivated by a number of studies arguing that a devaluation was likely to reduce the gross domestic product further and bring more inflation and a larger fiscal deficit (see, for example, BCN, *Simulación ilustrativa de una devaluación del córdoba* [Managua: BCN, 1982]). See also Pizarro, an economist of the Nicaraguan Ministry of Foreign Trade, "New Economic Policy," p. 217.

tion to the dollar inhibited export production, kept up the pressure for foreign exchange, and stimulated a contraband trade in goods for export and in products imported at the official exchange rate. Simultaneously, this overvaluation and the difference between the rate of exchange for exports and that for imports resulted in financial losses for the Central Bank that were covered with more paper-money issues. . . . In reality, neither the generous credit policy (with negative interest rates and 100% financing) nor the low price (owing to the exchange rate) for imported supplies was able to arrest the decline in profits from production for export that originated in the overvaluation. Setting guaranteed prices for producers as a compensatory mechanism was not enough to stimulate production for exports, either because those prices did not adequately reflect costs and profitability or because they permitted only the most efficient producers to profit, or simply because the prices were not announced in a timely way. The decision to issue Certificates of Exchange Availability [CDDs] in February 1982 in order to stimulate exports was never effective because the government did not have enough foreign currency to assign to producers.[103]

In the same vein, the 1985 National Economic Plan, which was also issued around the same time as the currency devaluation announcement, acknowledged that the exchange rate had been systematically overvalued and that farm-gate prices had often been lower than production costs.[104]

A few months later a study by CIERA, the Ministry of Agriculture's research center, found that "since the triumph of the revolution . . . we have experienced a reversal of the domestic terms of trade of agriculture, as we witness an increasing transfer of resources from the farm sector to the rest of the economy."[105] The amount of resources that an overvalued exchange rate had extracted from cotton alone between 1980 and 1984 was estimated by MIDINRA's analysts to be 1.3 billion cordobas.[106]

In 1987, as the government grappled with hyperinflation, an assessment of past price policies drafted by MIDINRA pointed out that:

> it is useful to identify the characteristics of the iterative process involving inflation, costs, support price, and exchange rates [between 1979 and 1985]. In [the crop year] 1981–82 inflation reduced the profitability of

103. Ibid. The article is the English updated version of the original Spanish report issued in March 1985 and published in 1986 as "La nueva política económica de Nicaragua: Un reajuste necesario," in *Economía de América Latina*, no. 14 (1986): 75–91.
104. SPP, *Plan económico 1985*, pp. 7–8.
105. Molina, Morales, and Neira, *Los términos económicos*, p. 5.
106. Interview with a MIDINRA manager, December 1985.

[farm] export products. As a result, in 1982 a selective devaluation [of the cordoba] was introduced through a system of multiple export exchange rates and [cost-plus] guaranteed prices to producers. Since the weighted average export exchange rate proved to be greater than the weighted average import exchange rate, the result was a growing exchange loss for the Central Bank which, together with the fiscal deficit and the expansion of domestic credit, increased the money supply and the inflation rate. The increase in inflation affected in turn the costs of production, which led to new increases in support prices for producers, and new implicit exchange rates. This led again to more exchange losses for the Central Bank, and compelled the banks to increase nominal credit in order to finance higher production costs. The outcome was more inflation, new import exchange rates, and new prices to producers. . . . It is important to underline that those adjustments to the nominal value of producers' prices were smaller than the rate of inflation, and implied in real terms a revaluation of the cordoba and a decrease in farm-gate prices. . . . Between 1980 and 1985 consumer prices increased 877 percent, while the weighted average exchange rate increased only 93 percent. . . . Farm gate prices of the crops that were produced for domestic consumption increased faster than farm gate prices of export crops. . . . The overvalued exchange rate and the system of multiple exchange rates distorted the relationship between costs and revenues among agricultural production units. As a result, the financial statements of the [agribusiness] enterprises did not reflect their true economic profitability, nor were they a reflection of the comparative advantage among different products.[107]

By 1985, then, it was becoming increasingly apparent that the financial performance of many state-owned agribusiness enterprises had been substantially affected by a macro price policy over which state managers had little control. Given the seriousness of the problem, what strategy did state agribusiness enterprises develop between 1980 and 1985 in order to survive financially? And what impact did their survival strategies have on the implementation of food policy?

We know that in the particular case of the State Cotton Farm (SCF), the outcome eventually was a reversal of its policy of increasing cotton production. For SCF, an alternative such as growing sorghum for domestic consumption became more profitable and increasingly compelling as losses and debts kept accumulating. Switching from cotton to sorghum was not, in itself, a novelty. Many private cotton producers had already reluctantly followed that path, stating that a policy that led to significant reductions in cotton production was

---

107. MIDINRA, *La política de precios agropecuarios*, pp. 2–11.

detrimental to their own interests, as well as to the food system.[108]

The shift away from export production toward production for the domestic market had occurred because of the declining profitability of the former as compared with nontradables. According to MIDINRA, between 1979 and 1985, farm-gate prices of products for domestic consumption had increased much faster than equivalent prices for export products.[109] Prices of beans and corn, for example, increased by a factor of 83 and 175, respectively, in that period, whereas prices of export crops such as sugar, coffee, and cotton had, by comparison, risen only 5, 26, and 39 times, respectively.[110] Yet significant questions remained: How generalized was this movement away from agroexport among state-owned enterprises? What changes did it bring to those state agribusiness firms that did not have the option to switch markets? To answer these questions, I turn in the next chapter to the study of the domestic market and to an analysis of the impact of domestic price policy on the food system.

Conclusion

The empirical evidence presented in this chapter reveals that the exchange rate was significantly overvalued in the cotton commodity system. Estimated shadow exchange rates diverged increasingly from the official exchange rate after 1979. By 1984, estimates of the shadow exchange rate had reached as much as twelve times the official value.

Loyally implementing government policy, but going against the actual national trend, state-owned enterprises increased their cotton production, only to become some of the biggest money losers of the APP state sector. In general, the overvaluation of the cordoba had a substantial negative impact on the financial results of state-owned enterprises, as the majority of them operated in the agricultural export sector. For SCF, this impact meant large financial losses. The country's overall cotton output, meanwhile, remained at about half the prerevolutionary level and, on the whole, clearly did not follow the growth path of the state sector. This fall in output was, in turn, responsible for the country's single largest amount of foregone export earnings since 1980.

108. See n. 65 of this chapter.
109. MIDINRA, *La política de precios agropecuarios*, p. 14.
110. Ibid., p. 15.

The decision to uncouple farm prices from the official exchange rate contributed to the development of a practice among food policymakers of ignoring the dollar price at which Nicaraguan exports were being sold. A complex cost-plus farm-gate pricing system, disconnected from the exchange rate, created the false impression among policymakers that export producers had a fair chance to make a profit. Under the CDD system of export incentives, neither the government nor farm producers knew the implicit rate of exchange of their export crop until eighteen months after the initial cost-plus price had been announced. The differentiated exchange rate system that resulted from these pricing mechanisms led to the simultaneous use of four government exchange rates: the official, the parallel, the implicit, and the multiple rates. The complexity of the system made it extremely difficult to assess its net effect in a timely way, thus contributing to the perpetuation of agricultural disincentives in the export sector.

4

Food Prices

One of the structural factors that prevented the development of the Nicaraguan food system [before 1979] was the presence of a disarticulated economy where a limited industrial base was incapable of processing most farm output. The Nicaraguan [postrevolutionary] development model is based on the expansion of agroindustry. By that it is meant an industrial sector which grows not separate from, but integrated with agriculture. Our industrialization is the industrialization of agriculture.
—Nicaraguan Ministry of Agriculture and Land Reform

The entire food-processing and agroindustrial sector is working at less than 50 percent capacity because farmers won't deliver their products to us. They say that our prices are too low and that in order to survive they have to sell in the black market. So, we do one of three things: close the processing plants, import the raw materials to keep them going, or join the black market. All three actions are in conflict with our stated policy.
—Director, state-owned food-processing plant

The financial difficulties of state-owned agribusiness enterprises not only impeded the food system's self-sufficiency, they also threatened the viability of another critical food strategy component: the creation of a new integrated economic development model with agroindustry as its foundation.

Government reports indicated that after the 1979 revolution, the entire agroindustrial sector, approximately half of which was state-owned, had been operating at less than 50 percent capacity. Major food-processing industries that were wholly state-owned, such as milk-pasteurizing and meat-packing plants, were operating at about 20 to 30 percent of capacity. This situation was particularly disturbing to government food policy analysts because of the financial burdens and food shortages it created. In addition, it encouraged state-owned food-processing plants to purchase foreign-subsidized unprocessed farm products instead of similar domestic products, at the expense of domestic producers and foreign exchange reserves. The presence of considerable idle processing capacity also prompted sales of processed output in the domestic black market, at the expense of low-income consumers and export growth.

This chapter examines the lack of coherence in the government's domestic food price policy, particularly with regard to the relationship between production and distribution costs, farm-gate prices, and retail prices, and the inadequacy of the government's price-setting and price-adjustment mechanisms devised to implement the policy. The chapter explores the possibility that this incoherence was responsible for many inefficiencies originating at state-owned processing stages of the food system and that price policies and their implementation mechanisms motivated a series of state enterprises' actions in contradiction with food policy objectives. These actions, in turn, contributed to the creation of food shortages, the development of an urban black market for food, and a deterioration of the rural-urban terms of trade.

Agroindustry as a Development Model: Goal and Performance

Many developing countries have instituted costly economic policies favoring industrial production and urban consumption at the expense of their predominantly agrarian and rural economies. Postrevolutionary Nicaragua, in contrast, based its economic strategy on a clear understanding that the nation's wealth depended on the fate of its farm sector.[1] The primacy accorded to agriculture determined the government's attitude toward the industrial sector and served as the basis of the country's industrial policy.[2] In the words of MIDINRA minister Jaime Wheelock:

> None of the development strategies that the Somoza regime and local capitalism implemented in Nicaragua ... during various stages of our history

---

[1]. See, for example, Junta of the Government for National Reconstruction, *Primer proclama del Gobierno de Reconstrucción Nacional, algún lugar de Nicaragua* (Managua: JGRN, June 18, 1979), p. 7; and Jaime Wheelock Román, *Entre la crisis y la agresión: La reforma agraria sandinista* (Managua: MIDINRA, 1984).

[2]. See MIDINRA, *Marco prospectivo del desarrollo agroindustrial*, vols. 1, 2 (Managua: MIDINRA, 1985).

were able to respond to the needs of the Nicaraguan people. Crisis, debt, hunger, inequality, and above all a significant [economic] backwardness are the result. After the triumph [of the 1979 revolution] we could not accept the perpetuation of a [development] model that was based [simply] on the export of agricultural raw materials, nor one of industrialization based on imported intermediate goods, because neither model was likely to get the country out of foreign indebtedness. . . . Nicaragua must base its development in the industrial transformation of its own natural resources, which are primarily in agriculture. Agroindustry must be the pillar of this industrial transformation which must also include the resources of forestry, fishing, and mining. . . . [Simply] to produce raw materials such as coffee, cotton, or wood does not allow for the development of other sectors of our society. . . . Instead of selling logs, we should sell finished wood products that we are now importing; instead of selling cotton, we should transform it and sell it as fabric or garments. And the same applies to leather and any other [farm] product. . . . In other words, it is from the [lines of activity of the] primary sector that we will develop industries. . . . As a result, agriculture has a central role to play in this process of industrialization."[3]

With this perspective in mind, the government made a preliminary assessment of the Nicaraguan industrial sector inherited with the downfall of the Somoza regime and issued its first comprehensive study of the country's food strategy.[4] It stated that a fundamental factor preventing the harmonious development of the Nicaraguan food system prior to 1979 had been "the presence of a disarticulated economy where a limited industrial base was incapable of processing most farm output."[5]

A subsequent, much more detailed study of the Nicaraguan industrial sector was conducted by MIDINRA in collaboration with the Ministry of Industry and the United Nation's Food and Agriculture Organization (FAO). This study partially confirmed the earlier assessment,[6] stating that before 1979, industrial activities centered around the processing of farm output had grown in a haphazard way, "without considering its linkages with other sectors of the economy and its multiplier effects. Therefore, never [before the revolution] had there been some strategic planning with regard to the development of agroindustry as such."[7]

    3. See Wheelock Román, *Entre la crisis y la agresión*, pp. 31–39.
    4. CIERA/PAN/CIDA, *Informe final del proyecto de estrategia alimentaria*, 4 vols. (Managua: MIDINRA, 1984).
    5. Ibid., p. 10.
    6. MIDINRA, *Marco prospectivo del desarrollo agroindustrial*, vol. 1, p. 11.
    7. MIDINRA, *Marco prospectivo del desarrollo agroindustrial: Resumen* (Managua:

The report also proceeded to conceptualize the history of the Nicaraguan industrialization process on the basis of Kalecky's model of a three-sector economy: one producing capital goods, another producing luxury consumer goods, and a third producing consumer goods accessible to medium- and low-income segments of the population.[8] Using this framework, the study concluded that in a small, underdeveloped country such as Nicaragua, where the potential for agricultural development is nevertheless relatively high, integration between the three sectors does not occur.[9] Instead, private entrepreneurs limit themselves to producing and exporting primary agricultural products, leaving the production of low-income consumer goods to a fragmented assortment of small producers operating with low capital-intensive technologies and minimal profit margins. Entrepreneurs then apply their foreign exchange earnings toward purchases of imported capital and luxury goods. Under these conditions, the report stated, the process of industrialization occurs in a fragmented and unplanned way. Subject to the vagaries of spontaneous private initiative, external pressures, and the opportunities of the moment, such industrialization eventually gives rise to a dualistic economy whose industry is highly dependent on the availability of imported intermediate goods and is deprived of a dynamic of its own.[10]

In keeping with the agroindustrial development model suggested by Wheelock, the report asserted that the industrial policy of postrevolutionary Nicaragua should not perpetuate this pattern. Instead, it suggested that policy ought to promote an industrial sector that is highly integrated with agriculture and therefore not dependent on imported foreign inputs for growth. As the MIDINRA study noted, "If this industrialization process occurs concomitantly with a gradual reduction in its imported components, it means that the Nicaraguan economy is developing in a manner consistent with its development model, which is a model of *industrialization of agriculture*."[11]

MIDINRA, 1985), p. 3; and MIDINRA, *Marco prospectivo del desarrollo agroindustrial*, vol. 1 (Managua: MIDINRA, 1985), p. 8.

8. MIDINRA, *Marco prospectivo del desarrollo agroindustrial: Resumen*, pp. 1–5.

9. Ibid.

10. Ibid.; see also Wheelock Román, *Entre la crisis y la agresión*, pp. 31–39.

11. MIDINRA, *Marco prospectivo del desarrollo agroindustrial: Resumen*, p. 1 (emphasis added).

The study also identified a major problem, however. Between 1979 and 1983 the Nicaraguan agroindustrial sector, which was characterized in the 1984 food strategy report as too small to process most farm output, had actually operated at less than half capacity.[12] Disaggregated data developed by the study covered forty-six different agroindustrial activities, of which thirty-six consisted of food-processing operations. Of the latter, fourteen were over 50 percent idle. Among the most important state-owned activities affected were slaughterhouses (53 percent idle capacity); sugar mills (49 percent); processing plants for meat (83 percent), milk (pasteurization, 65 percent, and powder, 57 percent), and cheese (40 percent); animal feed production plants (70 percent); fruit-canning factories (94 percent); and vegetable oil refineries (42 percent) (see Table 13).

The MIDINRA/FAO study stated: "The growth potential that can be derived from the use of [this] idle capacity is obvious. It is also obvious that the existence of this growth potential should give priority to actions that can expand raw materials supply [from the farm sector]."[13] It estimated that an efficient use of available plant capacity in the agroindustrial sector could increase the value of agribusiness processing by 83 percent without requiring additional investments in the immediate future.[14]

The report clarified that agroindustry was already the country's major industry, representing 75 percent of the entire industrial sector by output value and 15 percent of the gross domestic product. Within agroindustry, the most important component was food processing, its output value being approximately 60 percent of that of agroindustry as a whole. The remainder was specialized in the manufacturing of textiles, soap, leather, shoes, and wood furniture. Agroindustrial exports represented about 70 percent of the country's total exports. Of agroindustrial exports, about 90 percent consisted of refrigerated beef, processed coffee, semirefined sugar, ginned cotton, and powdered milk. The importance of state-owned agroindustrial enterprises was considerable, representing in 1983 an estimated 45 percent of the agroindustrial sector by output value.[15] Their presence within the food-processing industry was higher but varied greatly depending on the specific product (estimates are listed in Table 13). According to the MIDINRA/FAO estimates, idle plant

12. MIDINRA, *Marco prospectivo del desarrollo agroindustrial*, vol. 1, pp. 133–134.
13. Ibid.
14. MIDINRA, *Marco prospectivo del desarrollo agroindustrial: Resumen*, p. 31.
15. Ibid., pp. 2–40.

Table 13. Idle capacity of food-processing plants, 1983

Activity	Existing plant capacity (000 units)	Percentage state-owned	Percentage idle
Cattle slaughterhouses[a]	481 head	100	53
Poultry	10,230 head	30	10
Hogs	390 head	100	23
Meat processing	5,257 lbs	100	83
Milk pasteurization[b]	57,648 gals	100	65
Meat by-products	460 cwt	100	76
Industrial cheese	2,500 lbs	100	40
Powdered milk	25,000 lbs	N.A.	57
Tomato paste	168 cwt	100	28
Canned fruit	140 cwt	100	94
Vegetable oil	12,860 gals	38	33
Vegetable oil refining	15,130 gals	38	42
Rice milling	5,246 cwt	35	31
Wheat floor	1,398 cwt	N.A.	48
Coffee processing	2,000 cwt	37	22
Sugar milling	5,400 mt	65	49
Candles	7,000 lbs	N.A.	61
Ginger processing	436 cwt	N.A.	40
Margarine	5,500 lbs	N.A.	7
Cereal processing	14,700 lbs	N.A.	60
Coffee roasting	18,000 lbs	37	42
Instant coffee	4,000 lbs	N.A.	34
Vegetable fats	160 cwt	N.A.	52
Animal feed	10,200 cwt	60	70
Rum	4,200 gals	N.A.	42
Beer	23,640 gals	N.A.	42
Soft drinks	16,800 cases	80	42
Cotton ginning[c]	3,073 cwt	30	39
Banana processing	8,190 boxes	100	48
Egg production[d]	27,800 dozen	21	38
Poultry production[d]	11,412 head	32	29
Cassava by-products	13 cwt	N.A.	63
Bread[d]	160,000 lbs	N.A.	25
Cookies	23,300 lbs	N.A.	69
Pasta	12,000 lbs	N.A.	85
Egg incubation	23,750 hens	N.A.	51

Sources: Percentage state-owned: CIERA/PAN/CIDA and MIDINRA; and percentage idle: MIDINRA, *Marco prospectivo del desarrollo agronindustrial*, vol. 1 (Managua: MIDINRA, 1985), pp. 132–133.
[a]Includes national (export) and municipal slaughterhouses.
[b]Includes milk recombined from milk powder and vegetable fats.
[c]Includes cottonseed for cooking oil and animal feed.
[d]Includes industrial plants only.

capacity was highest in those food-processing segments with the greater number of state-owned plants.[16]

More specialized reports following the MIDINRA/FAO study

16. MIDINRA, *Marco prospectivo del desarrollo agroindustrial*, vol. 1, pp. 132–134; and CIERA/PAN/CIDA, *Informe final del proyecto de estrategia alimentaria*, vol. 4.

established that during 1983–85, the underutilization of plant capacity of state-owned food-processing plants was actually increasing. In 1985, for example, idle capacity had reached 63 percent for meat-packing plants, 73 percent for slaughterhouses in general, 81 percent for milk-pasteurizing plants, 78 percent for powdered-milk plants, and 56 percent for sugar mills.[17] By 1985, the low efficiency levels of state-owned processing plants had become a major government concern. Food producers were largely unresponsive, and food shortages and growing black market speculation were increasingly evident, particularly in the politically sensitive area of pasteurized milk and beef, two products that in the local culture were strong symbols of children's health and diet improvement. These developments prompted a high-level government decision in 1986 to conduct a formal investigation. Its primary focus was the problems affecting the eleven state-owned pasteurizing and packing plants since 1979 and, by extension, the evolution of the milk and beef agribusiness commodity system.[18]

Idle Capacity in Food Processing: The Cases of ENILAC and ENAMARA

The Growth of the Milk- and Beef-Processing Industry before 1979

The government's assessment of the food system inherited at the downfall of the Somoza regime (reviewed in Chapter 2) had identified the rapid expansion of the beef and milk commodity system between 1965 and 1979 as the most recent in a series of export-led booms in the prerevolutionary Nicaraguan economic growth pattern.[19] Between 1960 and 1979, the area under pasture had more

17. MIDINRA, *La ganadería en Nicaragua y sus perspectivas* (Managua: MIDINRA, 1986); and MIDINRA, *Perfil estratégico para desarrollar la agroindustria cañera* (Managua: MIDINRA, 1987), p. 63.
18. Interview, August 1986.
19. I conducted research primarily between May 1986 and July 1987 concerning ENILAC (Enterprise of the Milk Industry), ENAMARA (National Enterprise of Land Reform Slaughterhouses), and the problems of the milk and beef agribusiness commodity system between 1979 and 1985. I carried out part of this effort within the INCAE/Ford Foundation research project that I directed between 1984 and 1987, in collaboration with MIDINRA. I am particularly grateful to Mario de Franco for his invaluable insight and research that contributed toward an improved understanding of the milk and beef commodity system after 1979.
One of the many obstacles encountered in the effort to understand the dynamic of the milk and beef commodity system after 1979 was the inconsistencies in existing govern-

than doubled, reaching 11.6 million acres; the amount of cattle slaughtered had more than trebled, increasing from 133,500 head in 1960 to 465,500 head in 1979; the value of beef exports had jumped from less than $3 million to $94 million.[20] Domestic per capita apparent consumption of beef had also increased, from an average of twenty-one pounds in the early 1970s to slightly over thirty pounds in the second half of the decade.[21]

The expansion of the Nicaraguan beef commodity system was attributed to a number of factors. On the demand side was a growing need for cheap beef created by the expanding fast-food industry in the United States. On the supply side, Nicaragua had some real cost advantages in the production of grass-fed beef. The link between the pastures of Nicaragua and the world beef market was forged in 1957, when IFAGAN, the first modern packing plant in Central America approved by the USDA, was built in Managua by the Nicaraguan Institute of National Development (INFONAC) and the Nicaraguan Federation of Cattle Rancher Associations (FAGANIC). This link was further strengthened in the early 1960s by a breakthrough in refrigerator transport that reduced the cost of transportation and the risk of heat exposure. Traditional beef production, with its criollo breeding, crude pasture management, poor animal hygiene, wasteful transport to slaughter, unsanitary slaughterhouses, and medieval marketing networks, was replaced by the breeding of beefy Brahman herds, fenced rotational grazing, high-yielding grasses, chemical fertilizers, weed control, artificial insemination, and improved animal care, supported by specialized technical and credit assistance.[22]

---

ment data largely due to the fact that government statistics used many different units of measurement for the same product and frequently did not specify the unit of measurement used in a particular table or chart, creating the possibility of frequent confusions. For example, beef prices could refer to about ten different possible weights. Prices to producers might refer to the on-the-hoof head of cattle, before feeding or after feeding. This price per head referred to a standard steer of specific weight, but standards varied, weights varied accordingly, and prices became meaningless unless referred to the proper weight of the animal considered. Similar confusion existed regarding producer prices, wholesale prices, and retail prices.

20. FIDA, *Informe de la Misión Especial de Programación a Nicaragua* (Rome: FIDA, 1980), pp. 1–64; MIDINRA, *Marco estratégico del desarrollo agropecuario* (Managua: MIDINRA, 1983), vol. 1, pp. 1–7; vol. 2, pp. 1–23; Jaime M. Biderman, "Class Structure, the State and Capitalist Development in Nicaraguan Agriculture" (Ph.D. diss., University of California, Berkeley, 1982), pp. 111–127; and MIDINRA, *La ganadería en Nicaragua*, pp. 1–9.

21. CIERA/PAN/CIDA, *Informe final del proyecto de estrategia alimentaria*, vol. 4, p. 97.

22. Robert G. Williams, *Export Agriculture and the Crisis in Central America* (Chapel

The success of IFAGAN was followed by the construction of six other USDA-approved packing plants: CARNIC and EMPANICSA in 1962, IGOSA in 1971, the Amerrisque and San Martín plants in 1977, and EPCA in 1978. These modern plants operated almost exclusively for the export market and did not compete directly with the traditional national network of 112 municipal slaughterhouses, which supplied local demand primarily with second-choice beef not meeting the USDA's export requirements. All seven modern packing plants were controlled by private interests, and four of them (IFAGAN, CARNIC, EMPANICSA, and Amerrisque) were partially or wholly owned by the Somoza business group. Producer prices for cattle delivered to the packing plants were tied to international beef prices, varying within a range of two cents per pound on the Chicago "Yellow Sheet." On average, plant capacity utilization during the prerevolutionary period was estimated to have ranged from 60 percent to 75 percent. Since Nicaragua's packing plants were not affected by the aftosa cattle disease quarantine, had access to the U.S. beef import quotas, and could transfer the price fluctuations of the Chicago "Yellow Sheet" to domestic producers, Nicaragua's packing plants operated very profitably during the 1960–79 period. Cattle ranchers, in turn, expanded their operations considerably because, in general, they had a competitive cost advantage in grass-fed cattle raising and had the option of either absorbing the international price fluctuations, waiting for better prices, or selling to the municipal slaughterhouses. There were several notable offshoots of the growth of a modern meat-packing industry. One was the creation of a beef by-products industry, which included six sausage plants, nine tanneries, and four manufactures of shoes. A second offshoot was the development of an animal-feed industry, including fifteen plants processing cotton cake and cotton meal for cattle feeding. A third was the growth of a soap industry that processed animal fats.[23]

Most cattle ranches did not specialize in beef production but pro-

---

Hill: University of North Carolina Press, 1986), pp. 77–151; Mario de Franco and Brizio Biondi-Morra, *La agroindustria de la carne en Nicaragua*, 1987 (Managua: INCAE, 1987); and Beatriz Joya, Mario de Franco, and Brizio Biondi-Morra, *La industria láctea de Nicaragua*, 1987 (Managua: INCAE, 1987).

23. MIDINRA, *La ganadería en Nicaragua*; and John Brohman, *The Declining Performance of the Matadero Sector in the Post-Triunfo Era* (Managua: INIES, 1985); John Brohman, *Growth and Stagnation in the Nicaraguan Dairy Industry* (Managua: INIES, 1985); John Brohman, *The Nicaraguan Dairy Industry in the Post-Triunfo Era: Production Increases but Declining Entrega and Growing Foreign Dependency* (Managua: INIES, 1985).

duced both beef and milk. As a result, an expansion in milk production occurred in parallel with that of the beef industry. Nicaragua's first milk-pasteurizing plant, La Salud, was built in Managua in 1949. It was not until 1959 that a second plant, La Perfecta, was constructed, followed shortly thereafter by the same expansion pattern that had affected the packing plants. Six new plants were built in the 1960s: La Buena, La Reina, La Completa, and La Exquisita, in Managua; El Hogar, in the city of León; and PROLACSA, in the city of Matagalpa. In contrast to the packing plants, which operated almost exclusively for the U.S. and world beef market, milk-processing plants sold their dairy products in the domestic market, with the exception of PROLACSA, a joint venture between Nestlé and several regional investors. PROLACSA produced powdered milk for export in the Central American Common Market (CACM).[24] In the 1970s, following the construction of two additional modern plants in Managua, the milk industry was restructured around six plants. The first, PROLACSA, continued to produce and export powdered milk from its processing facility in Matagalpa. The other five, La Salud, La Perfecta, La Completa, La Selecta, and El Eskimo, supplied the domestic market with pasteurized milk and other dairy products from their plants in Managua.[25]

Total unpasteurized raw fluid milk production more than doubled during this period, increasing from 57 million gallons in 1960 to 120 million gallons in 1978. Raw fluid milk purchased by processing plants increased almost fivefold, from 3 million gallons in 1960 to 14 million gallons in 1978. Between 1960 and 1979, Nicaragua went from being an importer of about $1 million per year of dairy products to becoming a net exporter of these products in the amount of about $13 million per year. Domestic per capita apparent consumption of fluid milk also increased, from approximately nine gallons to eleven gallons between 1970 and 1978.[26]

Before 1960, each pasteurizing plant set its purchase and selling prices independently, which generated vigorous price competition, benefiting both producers of raw fluid milk and consumers of pasteurized milk. With construction of the new plants, however, common pricing policies were set under the terms of the "milk pool" agreement (*pool lechero*). Although these policies varied somewhat between 1960 and 1979, in essence they were all designed around

24. Brohman, *Growth and Stagnation in the Nicaraguan Dairy Industry*.
25. Ibid.
26. CIERA/PAN/CIDA, *Informe final del proyecto de estrategia alimentaria*, vol. 4.

differentiated pricing based on the fluctuations of milk supplies with the dry and rainy seasons of the year. During the dry season, when grass and animal feed were scarce and expensive, the supply of raw fluid milk decreased and the plants paid a higher fixed price than in the winter. In turn, plant procurement during the winter, when there was an oversupply of raw fluid milk, was based on lower prices; purchases from producers in that season would be made on a quota system proportional to the amount of milk delivered during the dry season.[27]

During the 1960s and 1970s, retail milk prices were regulated by the government. A policy of low and stable retail prices for pasteurized milk was sometimes mitigated through negotiation with the milk-processing plants at times when their profitability or their ability to procure raw fluid milk was threatened by falling margins. On average, however, retail milk prices had increased by only 5 percent annually from 1960 to 1979. Accordingly, the profitability of milk-processing plants had been, on the whole, lower than that of the packing plants and at times had even been negative. Plant capacity utilization during the 1960s and 1970s was estimated at around 60 percent.[28]

The Restructuring of the Milk and Beef Industry after 1979

After the 1979 revolution, the milk- and beef-processing industry and the agribusiness commodity system in which it operated were subjected to a number of structural changes (see Figure 4). Among the most important modifications were, first, the nationalization of all seven meat-packing plants. In the second half of 1979, these facilities were converted into state-owned enterprises reporting to a single parent company, the National Enterprise of Land Reform Slaughterhouses (ENAMARA), which, in turn, was managed by MIDINRA.[29] Second, in accordance with the Slaughterhouses Plan (*Plan de Mataderos*), municipal slaughterhouses were closed down in 1981, and the traditional regional meat wholesale network was

---

27. Jaime Román, *La industria láctea en Nicaragua*, case written under the supervision of Professor James E. Austin (Managua: INCAE, 1971); Harry Downing, *Breve diagnóstico de la situación de la producción e industria láctea en Nicaragua* (Managua: MIDINRA, 1979); John Brohman, *Growth and Stagnation in the Nicaraguan Dairy Industry*, and *Nicaraguan Dairy Industry in the Post-Triunfo Era*.
28. Downing, *Producción e industria láctea en Nicaragua*.
29. ENAMARA, *Enfoque de la situación coyuntural de los mataderos estatales* (Managua: MIDINRA, 1981).

Figure 4. The structure of the Nicaraguan milk/beef commodity system: estimated state share of output, 1983

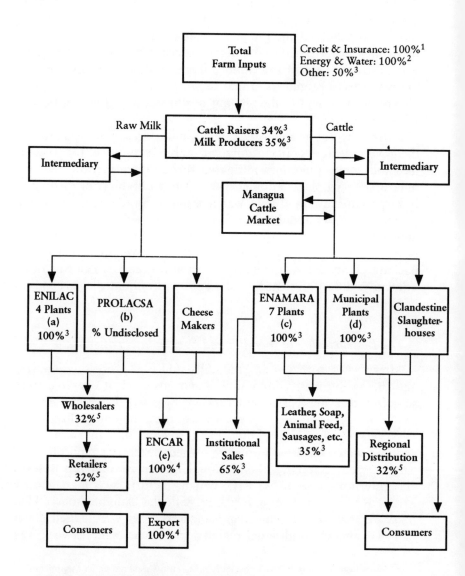

Sources: [1]BCN; [2]INE; [3]MIDINRA; [4]MICE; [5]MICOIN.
(a) Pasteurized milk and dairy products plants: SELECTA, PERFECTA, COMPLETA, and ESKIMO.
(b) Milk powder, partly state-owned.
(c) CARNIC, IFAGAN, ENPANICSA, Amerrisque, EPCA, IGOSA, and San Martin
(d) Semi-industrial slaughterhouses administered by municipalities.
(e) State-owned trading company that is the sole exporter of Nicaraguan beef.

eliminated, so as to give the state-owned packing plants and ENAMARA a complete monopoly over beef production and domestic distribution. Third, beef export operations were centralized and separated from the packing plants and ENAMARA. They became the sole responsibility of the Beef Enterprise (ENCAR), a new state-owned trading company created in September 1979 and reporting to the Ministry of Foreign Trade.[30]

Fourth, in 1983, MIDINRA implemented in the dairy industry the same process of state centralization it had carried out in the beef industry. In April of that year, the ministry created the Enterprise of the Milk Industry (ENILAC) as the state holding company in charge of supervising three pasteurizing plants, La Completa, La Perfecta, and El Eskimo, all of which had been nationalized shortly after the revolution, as well as the partly owned La Selecta and PROLACSA. ENILAC's formal objectives were to satisfy domestic consumer demand for pasteurized milk at stable prices and to implement incentive price systems that would stimulate milk producers to increase the delivery of raw fluid milk to the plants. To carry out these tasks, ENILAC was given a role in determining milk price policies and milk subsidies. Additionally, ENILAC had the authority to perform import operations, make investments, and provide technical assistance in the milk commodity system.[31] Fifth, cattle estates belonging to Somoza and his associates, as well as those expropriated under the 1981 Land Reform Law, were organized into twenty large state-owned ranches and diversified state farms, which, by 1983, owned about 35 percent of the country's pastures and livestock. Additional expropriated land was organized into cooperatives.[32]

The Crisis of ENILAC and ENAMARA

The primary motivation for the creation of ENILAC and ENAMARA as the two centralized state organizations controlling the country's milk and meat-packing plants was a desire to increase the effi-

---

30. MIDINRA, *La ganadería en Nicaragua*; de Franco and Biondi-Morra, *La agroindustria de la carne en Nicaragua*, 1987.
31. MIDINRA, *La ganadería en Nicaragua*; Downing, *Producción e industria láctea en Nicaragua*; Brohman, *Growth and Stagnation in the Nicaraguan Dairy Industry*, and *Nicaraguan Dairy Industry in the Post-Triunfo Era*; Joya, de Franco, and Biondi-Morra, *La industria láctea de Nicaragua*.
32. MIDINRA, *Uso actual de la tierra en las empresas de reforma agraria* (Managua: MIDINRA, 1984).

ciency and effectiveness of those two almost exclusively state-owned industries. Increased efficiency referred mainly to increased plant capacity utilization. MIDINRA removed the autonomy of individual packing-plant managers with regard to purchases of cattle and sales of beef; these purchase decisions became the responsibility of ENAMARA, as did the cost control function, since many packing-plant managers had no idea of their production costs. The same actions were taken with ENILAC and the pasteurizing plants. Because of the government's conviction that the plants' fixed costs were high, capacity utilization was thought to be a key determinant of the unit cost of output. The lower the unit costs, it reasoned, the higher margins would be. Greater margins did not need to be translated into increased profits for ENILAC or ENAMARA. They did, however, provide the financial basis for an incentive policy toward producers aimed at increasing the production and deliveries of raw milk and cattle to the plants. At the same time, municipal slaughterhouses were closed down and meat wholesalers were eliminated. The rationale behind these actions was the government's desire to concentrate the beef-processing function within the state-owned packing plants to increase their utilization and to eliminate the intermediaries that were thought to be making large profits at the expense of consumers.[33]

Increased effectiveness, therefore, meant basically two things: an improved ability to supply pasteurized milk and other dairy products to consumers at reasonable prices and the generation of increasing amounts of foreign exchange through the export of beef. In this milk and beef strategy, the government chose ENILAC and ENAMARA as the key implementation vehicles.[34]

Nevertheless, the results were opposite to what was intended. Idle capacity of meat-packing plants, which had averaged 30 percent between 1977 and 1979, increased to 85 percent in 1981 and averaged 60 percent in the following four years. This increase occurred despite the closing of the municipal slaughterhouses in 1981. The unintended outcome of this action had been a proliferation of clandestine slaughterhouses; the loss of most slaughtering by-products, such as hides and animal fats; and the emergence of an illegal network of beef distributors organized by the formerly licensed meat wholesalers whom ENAMARA had dismissed in its effort to establish its own meat distribution network.[35] The competition that this vertical-

33. Interview with the manager of a state processing plant, November 1986.
34. Ibid.
35. Interview, June 1987.

ly integrated black market system represented for ENAMARA became so strong that it prompted the government to reverse its beef-processing policy. After 1983 municipal slaughterhouses were gradually reopened, and some of the old wholesalers were reintegrated into the formal economy.[36] Clandestine slaughtering, however, continued to grow, and ENAMARA's idle capacity in 1984 and 1985 remained on average around 58 percent.[37] Similarly, idle capacity of milk-pasteurizing plants increased from 40 percent in 1977 to 80 percent in 1985. The unused capacity of the powdered-milk plant PROLACSA, which had been at a high of 63 percent in 1977 due to the doubling of its facilities in the preceding year, increased to 88 percent in 1980. Later it remained at around 80 percent. All pasteurizing and meat-packing plants reported substantial financial losses.[38] This fact was of particular concern to ENAMARA and ENILAC because packing plants had traditionally been highly profitable and pasteurizing plants received a government subsidy.[39]

Another cause of concern was the growth of subsidized imports of powdered milk and vegetable fats from the United Nations World Food Program (UNWFP) and the European Economic Community (EEC). Pasteurizing plants and, after 1982, ENILAC depended on these imports to produce recombined milk and to prevent an almost complete stoppage in the supply of processed milk to consumers. Except for the processing of recombined milk, the pasteurizing plants had been practically inactive since 1983, unable as they were to buy any significant quantity of raw fluid milk from producers.[40]

Data regarding foreign exchange generation were also unfavorable. Beef export volumes had decreased by 1981 to one-fourth their 1979 level, although export prices had remained basically unchanged.[41] During the subsequent period 1982–84, exports remained at about one-third their 1979 level, despite ENAMARA's centralization efforts. In 1985 they dropped to a new low, this time under the impact of the U.S. trade embargo beginning in May 1985.

Domestic beef consumption, on the other hand, had not been as

---

36. Interviews, June 1987.
37. MIDINRA, *La ganadería en Nicaragua*; and de Franco and Biondi-Morra, *La agroindustria de la carne en Nicaragua*.
38. MIDINRA, *Balances financieros* (Managua: ENAMARA, 1985); and MIDINRA, *Balances financieros* (Managua: ENILAC, 1985).
39. Interviews, June 1987; see also MIDINRA, *La ganadería en Nicaragua*.
40. Interview, June 1987.
41. See MIDINRA, *La ganadería en Nicaragua*.

affected as exports during the 1979–85 period. By 1984, the volume of domestic beef consumption had come to represent three times the volume of exports, an outcome that was inconsistent with food policy objectives. Including the amount of beef consumed through clandestine slaughterhouses, which was revealed in its full dimension by field research only two years after the fact and had not been accounted for in the official statistics, the domestic consumption of beef was probably greater than in the decade before 1979.[42]

Several other negative indicators apparently related to the previous results were additional sources of concern. First, an investigation by MIDINRA analysts in 1984 and 1985 showed that the actual number of head of cattle had dropped since 1979 to less than half the projected amount. Second, another MIDINRA investigation revealed a worsening balance of trade in beef (see Table 14). Increasing imports of inputs for cattle ranches and packing plants accompanied decreases in herd size, level of beef exports, and plant utilization. The amounts were so great that the positive beef export-import ratio had decreased from 46 in 1978 to 2 in 1985. Moreover, the value of imports per head of cattle had increased from $0.72 in 1978 to $4.64 in 1985 (Table 14). Third, milk production data revealed that the output of raw fluid milk had also dropped. Raw fluid milk production reached its lowest level in 1983, its estimated volume for that year being nearly one-fifth the 1978 level. After 1983 it began to increase, but in 1985 it was still at about one-third its prerevolutionary level. MIDINRA's preliminary investigations cited not only the fall in the total national herd size (steer and cows) as a likely cause but also a rise in the number of cows slaughtered, which implied a threat to the capacity for the national herd to reproduce itself.[43] Fourth, the milk- and beef-by-product industries had also been negatively affected. The resulting shortages of sausage, cheese, powdered milk, soap, and shoes fueled a growing black market and put additional pressure on the country's trade balance (see Table 15). Exports of hides, dairy products, hogs, and pork meat were virtually eliminated. In addition, the scarcity of milk and beef increased the country's dependency on imports of beef, powdered milk, poultry, and eggs and hindered the implementation of an export

---

42. Ibid.
43. Ariel Cajina, *El subsistema ganadero*, 4 vols. (Managua: MIDINRA, 1986); and MIDINRA, *La ganadería en Nicaragua*.

Table 14. Beef trade balance (US$ 000)

	1973	1974	1978	1984	1985
1. Total exports from cattle (beef, fats, hides, etc.)	54,138	46,174	89,399	19,800	13,500
2. Imports related to cattle (inputs for cattle farms and slaughterhouses, etc.)	1,448	997	1,952	4,402	6,495
3. Export-import ratio	37	46	46	4	2
4. Head of cattle (million)	2.4	2.5	2.7	1.6	1.4
5. Imports per head of cattle (US$)	0.60	0.40	0.72	2.75	4.64

Source: MIDINRA.

strategy aimed in part at diversifying the range of export products.[44]

In 1987, MIDINRA's new director of the livestock division described the sector's evolution during the 1979–85 period in the following terms: "We simply lost control of the beef and milk system. If we don't find a way to redress this situation in the immediate future, the damage will be felt for more than a decade."[45] Before a new milk and beef strategy could be designed, however, some basic questions needed to be answered. What had actually happened to the beef and milk commodity system? Why had the plants been unable to attract producers? Why had the milk plants resigned themselves to produce recombined milk from imported ingredients? Why was ENAMARA selling more beef in the domestic market than to the state beef-exporting company, ENCAR? Why had the national herd and raw fluid milk production decreased? Why had cattle ranchers slaughtered cows at their reproductive age? And why had clandestine slaughtering and the distribution of beef and dairy products in the black market spread so vigorously? Research conducted in 1986 and 1987 by MIDINRA, in collaboration with the INCAE/Ford Foundation research project that I directed, aimed to answer precisely those questions.[46]

---

44. Cajina, *El subsistema ganadero;* and Ministry of Foreign Trade, *Anuario de exportaciones, 1982* (Managua: MICE, 1982).
45. Interview, June 1987.
46. Two basic discussion papers, prepared in collaboration with MIDINRA, were Joya, de Franco, and Biondi-Morra, *La industria láctea de Nicaragua;* and de Franco and Biondi-Morra, *La agroindustria de la carne en Nicaragua.*

Table 15. Exports and imports of meat, poultry, and dairy products, 1977–1985

	1977	1978	1979	1980	1981	1982	1983	1984	1985
Exports									
Beef (US$ 000)	37,300	67,700	93,500	58,600	23,200	33,800	31,400	17,600	12,600
Hides (US$ 000)	3,800	5,600	3,300	2,500	1,500	400	2,400	—	—
Milk powder (US$ 000)	14,000	N.A.	—	—	—	22	—	—	—
Cheese	91	73	80	—	—	—	—	—	—
Eggs	—	36	41	—	—	—	—	—	—
Hogs and pork meat	2,638	2,418	1,129	29	—	—	—	—	—
Total exports	57,829	75,833	98,050	61,129	24,700	34,222	33,800	17,600	12,600
Imports									
Beef (US$ 000)	406	244	63	2	3,379	885	308	N.A.	N.A.
Hides (US$ 000)	359	352	581	2,242	6,642	9,093	7,584	N.A.	N.A.
Milk powder (US$ 000)	152	106	66	190	120	55	8	N.A.	N.A.
Cheese	896	683	845	5,610	6,716	1,014	2,114	N.A.	N.A.
Eggs	24	3	244	2,412	2,345	1,890	841	N.A.	N.A.
Hogs and pork meat	—	27	1	—	77	—	—	N.A.	N.A.
Total imports	1,837	1,415	1,800	10,456	19,279	12,937	10,855	N.A.	N.A.

Sources: CIERA/PAN/CIDA, *Informe final del proyecto de estrategia alimentaria*, vol. 4 (Managua: CIERA, 1984), for the period 1977–83; and MICE for the years 1984 and 1985.

The Impact of Domestic Food Price Policy

Field research conducted in 1986 and 1987 identified five basic factors responsible for most of the problems that ENAMARA and ENILAC experienced during the 1979–85 period and for the related unwanted food policy outcomes. These factors were: (1) the economic disruptions of 1979 due to the revolution, (2) the problems related to labor productivity since 1979, (3) the effects of the 1981 Land Reform Law and its related cooperative movement, (4) the effects of the counterrevolutionary war since late 1982, and (5) the impact of domestic food price policy during the entire 1979–85 period. Although the first four factors were important, the consensus among state-owned enterprise managers was that, on the whole, domestic food price policy was the key variable responsible for ENAMARA and ENILAC's performance.[47]

ENILAC and the Incentive to Increase Imports. Since ENILAC's consolidated financial statements were not up to date because of some accounting and financial control problems, one of the first tasks MIDINRA carried out in 1986 was to identify as accurately as possible the cost structure of its operations.[48] It was determined that in 1985, ENILAC lost 814 million cordobas on 1.3 billion cordobas in sales. In other words, losses represented 60 percent of sales revenues. This result was all the more disturbing because it occurred despite the milk subsidies received by individual plants and a very favorable interest rate on the debt. Without them, losses would have been even greater. More important, the variable costs of ENILAC were greater than its sales revenues, from which it follows that the contribution to fixed costs was negative. That fact had important implications. It proved what some state managers had always believed: that even if fixed costs were reduced to zero, ENILAC would still be losing money. The strategy of trying to increase capacity utilization so that plants would be profitable and would be able to pay better prices to producers was doomed to failure from the beginning. On the other hand, if the plants were to eliminate all purchases of domestically produced raw fluid milk and operate only on im-

---

47. Seventy-five out of eighty managers of state-owned enterprises that I interviewed stated that price policy was the main cause of ENILAC and ENAMARA's problems. The five who differed stated that the lack of farm inputs, such as animal feed, technical assistance, and the shortage of labor, were more important.

48. ENILAC carried out this task in collaboration with the INCAE/Ford Foundation research team between September 1986 and February 1987.

ported substitutes, they would become profitable. In 1985 alone they would have made a profit of 10 million cordobas, even assuming no cost savings or cost reductions other then those coming from a complete substitution of domestic milk purchases with cheaper imported substitutes. Under the same substitution assumption, profitability in 1984 would have been even higher, since the cordoba had not yet been devalued.

As a result, imports increased to such a degree that by 1983 the plants were almost completely dependent on them for operation. ENILAC was created in 1983 to reverse this trend. Domestic purchases had to be increased and dependency on foreign imports reduced. But the incentives to import were so great that the opposite happened: imports increased even more. The reasons for this increase become clearer if one considers that the government set the price paid to domestic producers, not ENILAC. Even though this price was much higher than that of its import equivalent, it was still too low for domestic producers. In 1984, 70 percent of the domestic raw milk delivered to ENILAC came from state-owned dairy farms because most private producers preferred to sell in the black market. In 1985 the participation of state-owned dairy farms increased to 90 percent. Almost no one else was delivering to ENILAC. By then, however, even state-owned dairy farms had begun to sell in the black market, just to be able to meet their payrolls.[49]

ENAMARA and the Incentive to Decrease Exports. To understand the problems of ENAMARA and its meat-packing plants, it is useful to see how the price ENAMARA paid for livestock was determined. Consider, for example, an actual increase made in 1985. The starting point for this calculation was the new 1985–86 retail price for one pound of beef, which the government increased from 62 cordobas to 459.82 cordobas (Table 16, line 12). There were essentially two reasons for increasing the retail price of beef: (1) to make beef more expensive relative to poultry, pork, and eggs so these products would become the main sources of protein and diet improvement and more beef could be exported and (2) to allow for an increase in producers' prices so cattle ranchers would be motivated to deliver their cattle to ENAMARA for slaughter and export.

ENAMARA, or one of its packing plants, would buy a standard 210-kilogram steer that slaughtered, cleaned, cut, and deboned

49. Interview, June 1987.

Table 16. Prices of milk and beef to producers, consumers, and exports, 1979–1986

		1979–80	1980–81	1981–82	1982–83	1983–84	1984–85	1985–86
	Official prices to producers							
1	Milk (C$/l)[a]	1.85	2.46	3.18	3.96	5.02	6.97	42.92
2	Milk (C$/l)[b]	1.2	1.61	2.40	2.78	3.95	6.45	36.86
3	Producer price index of milk[a]	100	133	172	214	271	376	2,320
4	Beef (C$/head of cattle)	2,446	3,328	4,071	5,086	5,294	14,780	17,470
5	Producer price index of beef	100	136	166	207	216	604	714
	Official and black market retail prices							
6	Milk, official price (C$/l)[c]	2.88	2.91	3.00	3.00	3.00	8.01	26.14
7	Consumer official price index of milk	100	101	104	104	104	278	907
8	Milk, estimated black market price (C$/l)	3.57	5.36	7.15	9.30	18	75	237
9	Consumer black market price index of milk	100	150	200	260	500	2,080	6,645
10	Beef, official price (C$/lb)	13	17	20	27	33	62	459
11	Consumer official price index of beef	100	130	153	207	253	477	3,530
12	Beef, estimated black market price (C$/lb)	13	30	40	53	63	153	1,511
13	Consumer black market price index of beef	100	233	311	409	482	1,179	11,628
14	General consumer price index	100	133	163	216	325	1,412	11,973
15	Milk consumer subsidy (C$ million)	N.A.	19	50	64	99	81	355[d]
16	Beef export prices (US$/lb)	1.20	1.30	1.11	1.08	0.95	0.90[e]	0.90[e]

Sources: MIDINRA, MICOIN, INEC, and MICE.

Note: C$/l = cordobas/liter.
[a] Price paid by pasteurizing plants.
[b] Price paid by powdered-milk plant, PROLACSA.
[c] Pasteurized milk, average of two years.
[d] Up to September 1986.
[e] U.S. trade embargo with Nicaragua begins in May 1985.

would yield between 310 and 334 pounds of beef. That beef could be sold in one of two ways. It could be exported, in which case it would be sold in cordobas to ENCAR (Figure 4), which in turn would sell it in dollars on the international market. Or it could be sold, also in cordobas, on the domestic market. Naturally, the government's policy was for ENAMARA to sell as much beef as possible to ENCAR. In that way, foreign exchange would be generated for the beef system and for the country.

The problems for ENAMARA began when the government-approved cordoba price of beef that ENCAR paid ENAMARA was not the same as the government-approved price received by ENAMARA for that same beef if it was sold on the *official* domestic market. ENCAR paid ENAMARA a cordoba price equal to the dollar price ENCAR received on the international market, converted in cordobas at the beef implicit exchange rate and less ENCAR's export commission. In this case ENCAR would pay ENAMARA 57,422.20 cordobas for the steer, or 171.92 cordobas per pound. ENAMARA, however, could sell in the domestic market for 459.82 cordobas per pound, which meant that every time ENAMARA sold to the export market through ENCAR, it lost 287.90 cordobas per pound. It also meant that if ENAMARA sold 100 percent of its output to the domestic legal market, its sales revenues would be 254 percent greater than if it sold all its output to the export market. ENAMARA's price incentives were precisely where they shouldn't be. At the time this example occurred, ENAMARA was selling 48 percent of its beef to the export market and 52 percent to the domestic market, so its total revenue from the sale of the slaughtered steer was 105,417.20 cordobas—the number of pounds of beef sold by the weighted average sale price that ENAMARA received from the sale of its beef. This weighted average price reflected the fact that part of the beef was sold to ENCAR and part to the domestic market. Once ENAMARA had determined that its sale revenues from that steer were 105,417.20 cordobas, it would deduct all its expenses, or 9,494 cordobas. These expenses were equal to ENAMARA's total costs divided by the estimated number of heads of cattle it would slaughter during the year. The difference between the revenues from sale of the beef and the costs allocated to that steer on a pro rata basis would determine the final price to the producer of the steer. In this case, the final price was 59,952 cordobas.

This pricing mechanism was the source of two problems. First, under this system all ENAMARA's costs, no matter how large, were always transferred to the producer. Therefore, from a strictly financial viewpoint, ENAMARA did not have much incentive to spread its costs over a greater number of head of cattle or to decrease its idle capacity. Under the circumstances, one may have expected ENAMARA to maintain poor control over costs. Since ENAMARA did not set its sale prices—these prices were set by the government—high packing-plant costs were likely to put pressure on the procurement function and result in low prices to producers. These low prices, in turn, were likely to result in a drop in the amount of cattle delivered to the packing plants. ENAMARA had been given a monopsony of the legal beef market, however. It was therefore assumed that cattle ranchers would, sooner or later, have to sell to ENAMARA, even if prices were not particularly remunerative.

This is exactly what happened, with one exception. ENAMARA did have poor control over costs. As a matter of fact, during the entire 1979–85 period, the packing plants and ENAMARA were run without cost control systems. Costs were estimated on the basis of a few rules of thumb, but no primary data collection system existed to make a precise cost calculation. Producers complained about ENAMARA's low prices and reduced their deliveries. But the exception to what would have been expected was that producers did have an alternative to selling to ENAMARA or holding on to their cattle: they could sell their products on the black market at higher prices than those offered by ENAMARA. Therefore, the prices ENAMARA offered had a negative impact on the delivery of cattle to the packing plants.

The decision made in 1981 to shut down the municipal slaughterhouses and to eliminate the regional wholesalers so deliveries to ENAMARA could be increased only served to reinforce the black market. The switch of the wholesale network to the black market was nearly complete, which was no small outcome: it involved a national network of people who knew the beef distribution business better than anyone else. Eventually, around 1983, ENAMARA reversed its decisions and tried to reenroll the wholesalers in the legal distribution channels. But by then the damage was done, and the black market had acquired a momentum of its own.

The pricing system engendered a second problem. As ENAMARA's production between 1979 and 1981 continued to fall, so did exports. ENCAR thus began pressuring ENAMARA to deliver more beef. But ENAMARA could do so only by paying higher prices to producers, which it could afford to do only if it raised its selling price of beef. Yet both the selling price to ENCAR and the domestic retail price of beef were beyond the state enterprise's control. The selling price to ENCAR depended on the Chicago "Yellow Sheet" and the beef implicit exchange rate. The domestic retail price was set by the government in coordination with the Ministry of Internal Trade (MICOIN). Ironically, therefore, the only way ENAMARA could raise money to increase its purchasing prices was to increase the share of its sales to the domestic market. Yet this action defeated its major objective, the increase of exports.

In terms of the actual numbers with which ENAMARA was dealing in 1985, its sales mix was then 48/52, that is, exports constituted 48 percent of total sales. This mix resulted in a producer's price of 95,923 cordobas for the 210-kilogram steer. Under this pricing mechanism, had ENAMARA's sales been 100 percent toward exports, a producer would have received only 51,658.20 cordobas for the steer, or 46 percent *less* than in the previous case. On the other hand, had ENAMARA sold all its beef to the domestic market, the same producer would have received 136,780 cordobas, or 42 percent *more* than under the 48/52 mix.

This explanation details what actually happened. After 1981, ENAMARA was able to reverse the fall in beef output by changing its sales mix. The price of this success, however, was the contradiction of the government's beef policy. Beef exports decreased, and domestic consumption increased. When municipal slaughterhouses, which served local markets, were allowed to open again to recapture the black market business, this trend expanded further.[50]

Producers and the Incentive to Join the Black Market. During the 1979–85 period, both state-owned and private cattle and dairy farms complained about high costs, low prices, and a worsening financial situation. The position of private producers regarding this period was summarized in two documents written in 1986. The first was a letter sent by FONDILAC, the association of private milk producers and processors, to the minister of MIDINRA.[51] The second was a letter sent by FAGANIC, the Nicaraguan Federation of Cattle Ranchers, to the president of Nicaragua.[52]

In its letter, FONDILAC addressed a single issue: Why did ENILAC and its state-owned milk-pasteurizing plants have problems in

---
50. Ibid.
51. Letter from FONDILAC to the minister of MIDINRA, September 16, 1986.
52. Letter from FAGANIC to the president of Nicaragua, October 31, 1986.

procuring raw fluid milk for their plants? According to FONDILAC, the primary causes involved ENILAC's operations and the problems faced by milk producers.[53] Both were related to price policies.

With regard to ENILAC, the document explained:

> The pasteurizing plants that are regulated and managed by ENILAC have suffered changes [after 1979] in their direction and in their objectives. What actually happens now is that the plants have their production of milk assured by imports of low-fat powdered milk and animal or vegetable fats which are either donated or purchased. Since the plants have their raw materials guaranteed and stored in their warehouses, they no longer have an interest in the milk produced in the country. In the past the directors of the milk-pasteurizing plants were in constant communication with and servicing milk producers. They constantly tried to attract producers who were selling to other plants. They also searched for new producers who were located in more distant areas. Obviously, this behavior was due to the fact that their only source of milk was the one produced within our country. Today the directors of the milk-pasteurizing plants think that they don't need domestic milk producers. They think that all that producers give them is problems, that they spend their lives fighting for milk quality, fat percentages, water percentages. Plant managers have said that they don't need milk producers any longer. If this is what plant directors think, the mentality of the rest of the plant personnel is the same. The attention given to milk producers is very bad, and milk producers have no place to go to complain. . . . As an example, a producer who has always delivered grade AR [top quality] milk to the plants recently received a cable from the plant saying that on his last delivery his quality came out to be lower [and therefore the purchasing price was reduced]. The cable also stated that if this decrease in quality happened again, they would discontinue plant purchases from him. In the past, the plant would have sent a technician to identify the cause of the reduction in quality. And current prices would have been maintained for two weeks, to give the producer a chance to correct the problem without incurring a penalty. In the past, milk quality tests were done by an independent laboratory. Today, no such independent laboratory exists. The laboratory belongs to the plant. . . . Since ENILAC owns the pasteurizing plants, the logical consequence is that ENILAC thinks as a pasteurizing plant. The plants pay according to the dictates of their own laboratories, and producers have no voice and no vote. In many cases this resulted in situations where a producer complains about something that he thinks is unfair, and the plant doesn't even listen to what he has to say, and in many cases the plant goes as far as discontinuing purchases from him. Another factor is the price of [raw fluid] milk. This price [after 1979] has always been controlled [by the government] and has not kept pace with the prices of farm inputs, which have increased

53. FONDILAC, pp. 1–5.

at a much faster rate. That forces many producers to abandon their activity as milk producers and to switch to other activities that are more profitable in the short run, such as the production of sorghum, corn, etc. At the same time, regional government administrators in most provincial cities allow for more frequent increases in the price of raw fluid milk than those of the plants. Therefore, we can see that in the towns of Masaya, Granada, and Nagarote raw fluid milk is sold in the official market at 800 cordobas per gallon, while the current price of the plants for the [best] AR milk is 435 cordobas.[54]

In its letter, FAGANIC, which comprised eighteen private associations of cattle raisers and ten thousand individual producers, also addressed a single issue: the crisis of cattle production and the measures needed to prevent its complete disappearance.[55]

The letter reviewed twenty-one pages of cost analysis covering every stage of cattle raising to demonstrate that with the cost-price relationships that had endured for the previous six years, cattle ranches could only lose money. As a result, explained FAGANIC, producers (1) did not replace their bulls when they died, or they replaced them with lower breeds, which meant future lower-quality herds; (2) did not maintain their ranches, so the number of head of cattle that one acre of land could sustain was decreasing; (3) were forced to sell cows that were in their reproductive years, thereby affecting the ability of the national herd to reproduce itself; and (4) were unable to keep their workers because of low salaries and poor living conditions, an occurrence that further damaged ranch operations and accelerated the exodus of the rural population to the cities.[56] An earlier cost-price analysis done by FAGANIC had already shown that between 1979 and 1984, the production costs of cattle ranches had increased by 433 percent, whereas ENAMARA's prices had increased by only 133 percent.

The situation of state-owned dairy farms and cattle ranches was essentially the same. Managers of these farms reported that the price of milk paid to them by ENILAC barely covered their variable costs. Some of these farms began selling half their production to the informal market, where prices were several times higher than the official price.[57] By 1985 the country's largest dairy farm, the state-owned company Roberto Alvarado, commonly known as Proyecto Chil-

---

54. Ibid.
55. FAGANIC, p. 1.
56. Ibid., pp. 8–10.
57. Interviews, September–December 1986.

tepe, supplied 90 percent of the raw fluid milk delivered to the pasteurizing plants. The development of this state-owned dairy farm had begun in 1981 in response to a fall in national milk production and the resultant retail shortages. According to its manager, however, "the price for delivering the milk to ENILAC amounted to financial suicide." The company, as a result, began negotiating the possibility of delivering only half its production to the plants.[58] State-owned cattle farms, unlike some of the modern state dairy farms, did not have cost control systems and could not make a direct comparison between costs and prices; however, they all reported overall financial losses.[59]

A comparative analysis of milk and beef prices revealed some clear patterns between the farm cycle years 1979–80 and 1985–86 (see Table 16). Although official producer prices during this period had increased 17 times for milk and 7 times for beef, the general consumer price index had increased 119 times. "Under these conditions," explained a state manager, "it is not surprising that cattle ranchers felt their entire life-style was being threatened."[60]

In addition, after 1980–81, milk prices paid to producers were higher than the retail price of pasteurized milk (Table 16, lines 2 and 8). "That fact," explained a milk plant manager, "was possible because of large milk subsidies. However, under those conditions retail shortages were almost inevitable. Theoretically, anybody that had access to pasteurized milk could resell it to the plant at a profit. In 1983–84 that profit was 67 percent of your investment, not a small thing if you could do that in a single day. Although this kind of 'recycling' phenomenon did occur with other products, as in the case of black beans for example, it did not happen with milk because you could resell your gallon of pasteurized milk on the black market at an even greater profit."[61] The estimated black market of price of milk in 1983–84 was about eighteen cordobas per liter, whereas the official price had remained at three cordobas per liter since 1980 (see Table 16).

In contrast with milk, there was no price disparity between consumer and producer beef prices because of the method by which beef was priced. The official retail price of beef in 1985–86, for example, was 459 cordobas (Table 16, line 12), that is, the same as

58. Interview, September 1986.
59. MIDINRA, *Reportes financieros* (Managua: MIDINRA, 1985).
60. Interview, June 1987.
61. Ibid.

the one used by ENAMARA to pay producers. Prices in the black market, however, were about two to three times those on the official retail market. Black market margins, therefore, were much greater, and the financial advantages associated with black market sales were obvious. In the opinion of one of MIDINRA's regional directors: "Some people thought that we had to eliminate beef wholesalers and intermediaries. Actually, the survival of the intermediaries was what saved the cattle sector, because those were the people that offered prices that were twice as much as ours and allowed [cattle ranchers] to survive. Had we succeeded in eliminating them, at this point all cattle ranchers would have been out of business."[62]

An estimate of the size of the black market for processed milk had not been made. Reports concerning these activities were denounced periodically in the government press, but the information was largely anecdotal and failed to provide quantitative estimates.[63] State plant managers were aware of the problem but encountered difficulties in dealing with what appeared to be a generalized practice. "The other day I was driving from Managua to Granada to visit a milk producer," explained one ENILAC manager, "and what do I see? Right there in the middle of the road, one of our refrigerated milk trucks had parked, and people were carrying all the milk containers away and reloading them into an unmarked truck. Just like that, in front of everybody. We cannot tolerate this kind of business. But what do you do? Do you fire all your truck drivers? Ironically, there is a high turnover among workers in the milk plants. But nobody wants to give up the trucking business. To be the chauffeur of a refrigerated milk truck has become more desirable than to be the plant manager."[64] Research showed that cheesemakers, operating primarily in the black market or in the uncontrolled informal economy, were ENILAC's prime competitors for producers' milk. The high prices at which they resold milk enabled them to offer better prices than did ENILAC and to attract official distributors of pasteurized milk.[65]

In the beef sector, comparative analysis of data coming from ENAMARA, the Nicaraguan Institute of Statistics and Census (INEC), and the Ministry of Planning indicated that clandestine

62. Interview, November 1986.
63. See *Barricada*, 1984 and 1985.
64. Interviews, November 1986 and June 1987.
65. See Joya, de Franco, and Biondi-Morra, *La industria láctea de Nicaragua*.

slaughterhouses had processed about 18,000 head of cattle in 1980 and 41,000 head in 1984. The latter figure represented about 3 percent of the estimated national herd. Other sources, however, suggested that the size of black market operations was much greater.[66] Discrepancies between data coming from INEC and MIDINRA appeared to substantiate that perception.[67] Another survival strategy of cattle ranchers was to smuggle their cattle to Honduras, a phenomenon that appeared to have continued after 1979. MIDINRA estimated that a steer priced at 96,000 cordobas on the official market was selling for 157,000 cordobas in the black market and for 325,000 cordobas at the border with Honduras.[68]

Another MIDINRA study showed that the price of beef had decreased markedly relative to other farm products (see Table 17). Particularly striking was the change in relative prices between milk and beef. One pound of beef bought six liters of milk in 1981 and fourteen liters in 1986. "A live cow is like a flow of small amounts of money over a long period of time," explained one of MIDINRA's regional directors, "while a slaughtered cow means lots of money in a single day. Why would someone who is in both the milk and beef business—as were most Nicaraguan cattle ranchers—prefer to kill the cow? In times of inflation, one would have assumed that the rancher would prefer to keep the cow alive, as a form of protection against inflation. But here they are killing the cows. If you look at the relative prices, you will find that, from the point of view of his own economic interests, he made the correct decision."[69]

Consumers: Squeezed between Food Shortages on the Official Market and Food Availability at High Prices on the Black Market. Despite the drop in domestic deliveries of raw fluid milk, pasteurizing plants had been able to supply processed milk in volumes larger than before 1979. That result, meant mainly to respond to consumer needs, was achieved by rapid increases in the production of recombined milk made with imported ingredients, while the supply of pasteurized milk had actually dropped to less than 15 percent of prerevolutionary levels. From the consumer viewpoint, the increases in processed milk output and low retail prices were obviously very pos-

66. See Brohman, *Matadero Sector in the Post-Triunfo Era;* and MIDINRA, *La ganadería en Nicaragua.*
67. Interviews, June 1987.
68. MIDINRA, *La ganadería en Nicaragua.*
69. Interview, November 1986.

Table 17. Purchasing power of one pound of beef, 1981 vs. 1986

Product	Unit	Amount of product whose retail value[a] is equivalent to 1 lb of beef	
		Dec. 1981	Sept. 1986
Tortilla	lb	5.2	0.6
Bread	lb	3.7	1.4
Cabbage	unit	3.3	1.9
Poultry	lb	2.0	2.0
Eggs	dozen	1.8	1.9
Pork	lb	0.9	0.9
Cheese	lb	1.7	1.4
Milk	liter	6.1	14.4

Source: MIDINRA elaboration based on data from INEC.
[a] At official (regulated) prices.

itive outcomes. Several factors, however, prevented consumers from fully benefiting from these results. First was the very low price at which pasteurized and processed milk were being sold on the official market. During the entire 1979–84 period, milk retail prices had remained practically unchanged at three cordobas per liter, while the general consumer price index had increased by 225 percent (Table 16). This policy of low milk prices had generated a very high demand, particularly from those low-income urban segments of the population that in the past had been unable to purchase pasteurized milk. This demand was thought to increase every year, as inflation rendered milk ever more inexpensive relative to other products. In light of the overall drop in the domestic production of milk, neither ENILAC nor local producers were in a position to satisfy the demand. Despite price and distribution controls, the disparity between demand and supply generated a rapidly expanding black market, where prices were sometimes as much as nine times the official price. Often milk at official prices was sought, not for consumption purposes, but for its resale value.[70]

Beef price differentials between official and black market networks were not as great as those for milk, the latter being about two to three times higher than the former (Table 16). "The reason why price differentials were higher in milk," explained a MIDINRA manager, "is because milk shortages were greater than beef shortages. The official retail price of beef had increased at a faster rate than the one for milk [Table 18], and therefore, the demand for beef

70. Interview, June 1987.

Table 18. Index of official and black market consumer prices of food (1980 = 100)

		1980	1981	1982	1983	1984	1985	1986
Index of official retail prices								
1	Corn	100	100	100	95	147	1,065	5,056
2	Rice	100	103	107	112	144	474	1,360
3	Beans	100	100	100	99	122	569	3,739
4	Sorghum	100	100	100	98	126	634	1,654
5	Milk	100	101	104	104	104	278	907
6	Beef	100	130	153	207	253	477	3,530
7	Broilers	100	114	136	163	274	974	N.A.
8	Eggs	100	99	113	136	271	778	N.A.
Index of estimated black market retail prices								
1	Corn	100	136	140	216	660	3,267	4,256
2	Rice	100	126	126	154	345	1,834	25,545
3	Beans	100	130	124	155	300	1,834	9,021
4	Sorghum	100	114	117	272	449	1,899	N.A.
5	Milk	100	150	200	260	500	2,080	6,645
6	Beef	100	233	311	409	482	1,179	11,628
7	Broilers	100	237	283	340	572	2,032	N.A.
8	Eggs	100	N.A.	N.A.	N.A.	N.A.	N.A.	N.A.
Consumer price index		100	133	163	216	325	1,412	11,973

Sources: For official prices, MICOIN, MIDINRA, INEC; for black market prices, estimates of INEC and MIDINRA; and for consumer price index, BCN.

had not increased as much as the demand for milk. At the same time, ENAMARA was increasing its sales in the domestic market, cattle ranchers were getting out of the milk business by converting their cows into beef, and the black market network for beef was much greater and better organized than the one that handled milk."[71]

The Missing Link: The Translation of Price Policy into Timely and Consistent Operational Decisions. In its analysis of the Nicaraguan agroindustrial sector, the 1985 MIDINRA/FAO study stated that before 1979, no comprehensive agroindustrial strategy had ever been developed, nor was there a public institutional framework for the formulation of an industrial policy. The same study concluded, however, that until 1985 the new government had achieved no progress either.

> Despite the fact that a number of research projects have been completed to face this issue, and despite the fact that recommendations have been made to institutionalize the harmonious development of agroindustry, the truth is that the lack of definition of an institutional framework still persists. The administration of state-owned agroindustries, and the direction and

71. Ibid.

supervision of the agroindustries pertaining to the private sector, constitutes a quite recent development within the [Nicaraguan] state. Various ministries and state agencies participate in this administration and supervision. It follows that planning, implementation, and control of operations are hindered by all kinds of adjustment problems which, in many instances, prevent the attainment of policy objectives.... The decision-making processes so far adopted always require an excessive number of state institutions, and decision making is extremely fragmented.[72]

The report then proceeded to enumerate the government institutions that had broad decision-making authority over the agroindustrial sector. The list included twenty-three major institutions and organizations belonging to MIDINRA, such as regional offices, central and general offices, PAN, AGROINRA, TECNOPLAN, and SUMAGRO; five organizations belonging to the Ministry of Industry; and fifteen other ministries and special organizational units, such as the Ministry of Internal Trade (MICOIN), the Ministry of Foreign Trade (MICE), ENAMARA, ENCAR, FNI, and the Ministry of Planning (MIPLAN). A complete listing would include nearly one hundred state institutions having some oversight of the agroindustrial sector. The report also stressed that individual agribusiness commodity systems were artificially broken down under different supervising ministries. Rice, for example, was under MIDINRA's responsibility only until the harvesting stage; milling and distribution were covered by the Ministry of Internal Trade and ENABAS. Cotton, on the other hand, was supervised by MIDINRA, both in the farming and the ginning stages, but cottonseed and cotton oil were under the Ministry of Industry (MIND). At the same time, new vegetable oil plants, such as those for coconut oil and African palm oil, were under MIDINRA management, which split the vegetable oil industry between two different ministries. One sausage plant belonged to the Ministry of Interior, another to the Ministry of Industry, and a third to MIDINRA. Processed fruit and vegetable plants were managed by MIDINRA, but the Ministry of Industry provided them with technical assistance and had decision authority over their imported inputs. A typical administrative decision regarding formulation of the annual plan for a line of activities, such as milk or beef processing, involved a minimum of thirteen different steps and twenty-five different organizations or committees.[73]

Decisions on prices and price policy for the beef- and milk-processing industry followed the same pattern, with pricing controlled

---

72. MIDINRA, *Marco prospectivo del desarrollo agroindustrial*, vol. 1.
73. Ibid.

Food Prices   131

by the government. Depending on whether it was an export price, retail price, plant purchasing price, or farm input price, however, the decision would be taken alternatively or jointly by the Ministry of Foreign Trade, the Central Bank, the Ministry of Internal Trade, and MIDINRA.[74] In the case of raw fluid milk purchased by the pasteurizing plants, for example, the process for establishing a new price was usually initiated by ENILAC. "After ENILAC suggests a new price," explained one of the plant managers, "the proposal goes to the General Cattle Division [GCD] of MIDINRA."

> After the General Cattle Division reviews, modifies, and finally approves the proposal, it goes to the General Economic Division [GED] of the same ministry. The General Economic Division, in turn, checks all the numbers and the cost analysis that supports them and introduces additional criteria that should be used in the final analysis. The modified proposal then goes to the office of the minister. The minister submits the proposal to the ministry's board for approval. If the board approves it, it goes to the National Planning Council [NPC] for the final decision. Every economic branch of the government has a member on the council. The council itself is headed by the president of Nicaragua. Once the council approves the proposal, the price change becomes effective. The proposal, of course, can go through this process several times before a final decision is reached by the council. The last time we made a price proposal, it took more than six months to get it approved. By then, the new price was inadequate due to its erosion by inflation."[75]

Added a state manager in a regional office:

> The length of this process contrasts with the speed at which local exceptions are granted once the official national price is approved. Usually this official price, which is normally established for the entire calendar or crop year, is enforced for only three or four months. Then a local government office, such as a regional office of the Ministry of Internal Trade, for example, grants a regional price rise because of special local circumstances. This price change is done without much consultation and often triggers all kind of demands from other regions. In the dairy region of Nagarote, for example, one such quick decision destroyed three months of patient work made by ENILAC to build a new network of milk suppliers. A sudden price rise approved in a contiguous area had made it more profitable to sell milk elsewhere. Such incidents were relatively frequent."[76]

---

74. Interviews, January–June 1987. See also MIDINRA, *Marco prospectivo del desarrollo agroindustrial*, vol. 1.
75. Interview, June 1985.
76. Interviews, November 1986. See also Brohman, *Matadero Sector in the Post-Triunfo Era*; and MIDINRA, *La ganadería en Nicaragua*.

Another problem was that the composition of a price-setting committee would vary with each farm product. The various government branches with direct responsibility over the farm sector and over the various stages of the food system, particularly the Ministries of Agriculture and Land Reform, Internal Trade, Foreign Trade, and Industry, were organized by product lines. MIDINRA, for example, would have a sugar department, a rice department, or a cotton department. The same would be true for the other ministries. A rice price-setting committee would have responsibility over only rice. Its membership would vary depending on whether the price to be defined was a farm-gate price, a milled price, a wholesale price, or a retail price. A committee on rice farm-gate prices would be composed predominantly by MIDINRA representatives of the rice department, and it was understood that MIDINRA, as the ministry overseeing food production, would have the final word. The same would apply for a rice retail price, except that in this case the chief role would be played by the Ministry of Internal Trade, the ministry overseeing food distribution. Frequently, a price committee meeting would be attended only by representatives of the predominant ministry because the other ministries felt that their presence would not make a difference in the final decision. Each ministry was known for being protective of what it considered its own turf. At the same time, most members of a rice committee would not participate in price committees of other products, and they knew little about the price decisions made with regard to other products, even when these products were potential substitutes. As a result, little overall domestic price policy coordination was achieved. The coordination problem existed in other policy areas, as well. In fact, the problem was so widespread that it became known as *feudalismo ministerial*, which indicated that despite the many interministerial committees, government branches behaved more like independent fiefs.

The objective of each individual price committee member also varied considerably, depending on a number of factors: the background and inclination of the person, the ministry to which the person belonged, the degree of control that the state had over a specific commodity system, the structure of producers and consumers of a particular commodity, and the circumstances of the moment. A representative of MIDINRA, particularly one who came from a family of farm producers, would, in general, have a better understanding and be more sensitive to the needs of producers than, for example, a representative of the Ministry of Internal Trade, whose main concern was typically the consumer. The former would be inclined to favor a price increase, whereas the latter would tend to oppose it. Sugar, a capital-intensive sector in which the state had invested large sums in mill expansions, would receive far more favorable price consideration than would coffee, a sector in which production was highly fragmented and state enterprises less prominent. Under the circumstances, the decision on the price of a particular commodity was made independently of the price of other products and was not guided by an integrated vision of the direction toward which prices should go. As a result, price relationships among different products could vary substantially, as could relative profitabilities, with the government becoming conscious of these variations and their implications only after their occurrence. This created unexpected shifts in production emphasis and substitution effects that changed the traditional dynamic of the commodity system and made its behavior highly unpredictable.[77] Typical was the change of emphasis in production from milk to beef when milk lost more than half its value in terms of beef (Table 17). Government policymakers did not intend that change in relative prices. It was an example of what unpublished government documents after 1984 referred to as "chaos in the pricing system." More complete comparative price analysis at the producers' level revealed other such cases (see Table 19). Between 1979–80 and 1985–86, the producer price of corn had increased 174 times, while rice had increased only 26 times and sorghum, 27 times. Partly as a result of that factor, many rice producers were switching to other basic grains, abandoning some of their previous investments in irrigated paddies.[78] In 1984–85, prices of farm export products such as cotton, sugarcane, coffee, and beef had increased less than those for crops used for domestic consumption, such as rice, corn, and sorghum (Table 19). Production of domestic crops was becoming even more attractive relative to export farming if domestic black market retail prices were used as the basis of comparison (Table 18). The switch from export farming to production for the domestic black market provided the highest price incentives and diverted substantial production capacity into new speculative activi-

77. See CIERA, *Los precios relativos* (Managua: MIDINRA, 1986).
78. Interviews, February–May 1986.

Table 19. Index of official prices to farm producers, 1979–1986 (1979–1980 = 100)

		1979–80	1980–81	1981–82	1982–83	1983–84	1984–85	1985–86
1	Corn	100	162	167	217	300	1,083	17,500
2	Rice	100	127	227	294	623	1,454	2,727
3	Beans	100	150	178	194	216	444	8,333
4	Sorghum	100	114	152	152	184	893	2,857
5	Milk	100	133	172	214	271	376	2,320
6	Beef	100	136	166	207	216	604	714
7	Broilers	100	173	177	189	301	496	2,554
8	Eggs	100	154	170	186	220	337	2,007
9	Cotton	100	110	133	139	194	340	2,387
10	Coffee	100	190	100	132	160	750	2,650
11	Ginger	100	131	131	197	210	1,190	5,460
12	Sugarcane	100	133	162	167	208	542	542

Source: MIDINRA.

ties.[79] Changes in the relationships between official and black market retail prices created other speculative opportunities for the alert trader (Table 18). Under those conditions, an increasing number of state-owned production enterprises, whose inputs were gradually approaching black market prices, reluctantly joined that same black market to be able to continue operating.[80] The overall assessment made in 1985 by the Ministry of Planning appeared to be highly appropriate: "The situation is one of chaos in the pricing system, which introduces irrationality in the allocation of products, discourages production, and promotes speculation and smuggling."[81]

Promoting Speculation and Smuggling: The Emergence of the Black Market

In 1987, as the government grappled with hyperinflation, an assessment of past price policies drafted by MIDINRA indicated that "between 1980 and 1985 economic policy favored the demand side and offered agricultural products at low prices. In this period, consumers were subsidized and agricultural producers' nominal prices were only moderately adjusted, so that farm prices decreased

---

79. Interview, June 1986.
80. Interviews, June 1987.
81. SPP, *Plan económico 1985* (Managua: SPP), pp. 1–3.

in real terms and the rural-urban terms of trade deteriorated."[82]

This phenomenon could be identified in a number of broad price patterns covering the 1979–85 period. In the specific case of the beef system, producers could demonstrate the deterioration of the rural-urban terms of trade by pointing out that the prices they paid for purchases from the urban and industrial sector—such as farm tools, professional services, mineral nutrients for their cattle, gas, and electricity—had increased between 3 and 7 times in five years, while the price of the farm products they sold had increased only 1.3 times. More generally, after 1982 the consumer price index had increased faster than prices of farm products (see Tables 18 and 19). Urban black market retail prices of food, on the other hand, had increased faster than the consumer price index (see Table 18). It appeared, therefore, that profit opportunities existed in the urban areas operating outside the formal economy and were beyond the reach of government price controls. The experience of ENILAC and ENAMARA had provided evidence to that effect, and the food policy implications of the simultaneous presence of idle plant capacity, supply shortages, and a vigorous black market caused concern at the highest levels of government.[83]

Moreover, government reports suggested that these problems were not confined to the milk and beef system. According to the Ministry of Planning, one of the symptoms of the generalized deterioration of the country's economic situation was "the flourishing of a speculative commercial sector which distorts the prices of products, removes huge quantities of cash from the National Financial System, attracts labor, and has come to represent a sort of parallel and parasitic economy which increases at the expense of what might be called the formal economy."[84] A substantial amount of the labor attracted to the black market economy was thought to have come from rural wage earners who were abandoning productive activities in the farm sector and migrating to the urban centers, primarily Managua.[85]

By 1985, government documents were reporting that the expansion of the black market threatened the formal economy and that domestic price policy was promoting speculation and smuggling.[86] Yet, before any attempt at redressing this situation could take place,

82. MIDINRA, *La política de precios agropecuarios* (Managua: MIDINRA, 1987).
83. Interview, May 1986.
84. SPP, *Plan económico 1985*, pp. 1–3.
85. MIDINRA, *Política de precios agropecuarios*; and interviews with MIDINRA managers, June 1987.
86. SPP, *Plan económico 1985*, pp. 1–3.

two fundamental questions needed to be answered. Why had the black market developed so rapidly in the first place? What were the basic forces fueling its growth? ENAMARA's experience with clandestine slaughterhouses proved that the problem was deeper and more difficult to resolve than originally suspected. In response to these questions I turn, in the next chapter, to the study of wage policies and to an analysis of their impact on labor productivity in the agricultural sector and the growth of the parallel, or black market, economy.

Conclusion

Field research on the milk and beef agribusiness commodity system reported in this chapter indicated that domestic food price policy was the primary cause of the considerable idle capacity in state-owned milk- and beef-processing plants. The fact that by 1985 these plants operated at about 20 to 30 percent capacity was related to a significant drop in deliveries of raw fluid milk and cattle to the plants, the proliferation of clandestine slaughterhouses, the corresponding loss of most slaughtering by-products, the emergence of an illegal network of milk and beef distributors, and a decrease in the national cattle herd.

Price incentives were found to have significantly contributed to the development of these problems. Price incentives were found to motivate cattle and dairy farms, both private and state-owned, to reduce their deliveries to the processing plants, sell their products in the black market, and switch production from milk to beef. Price incentives were also found to motivate milk-processing plants to increase imports of milk substitutes, neglect domestic milk producers, and sell processed milk in the black market. Finally, price incentives were found to motivate meat-packing plants to decrease exports, promote domestic beef consumption, and transfer to cattle ranchers the costs associated with their idle capacity. In this process, producers saw the prices of goods and services, including consumer products and farm inputs, increase several times faster than the prices of the milk and beef they sold to the processing plants; consumers also found themselves squeezed between food shortages in the official market and high food prices in the black market. All these actions and outcomes were in contradiction with food policy objectives.

The price policy decision-making process proved to be complex and time consuming. In the case of beef and milk, for example, price-setting procedures involved a minimum of four different ministries, seven administrative steps, and six months of negotiations among various government branches before a decision on a new price could be reached. This lengthiness and complexity prevented the translation of price policy into consistent and timely operational decisions and contributed to the perpetuation of perverse price incentives.

5

Wages

The participation of workers in state-owned enterprises . . . means the creation of a new labor discipline where . . . production is highly efficient.
—Minister of Planning, 1980

What we must solve in state enterprises is the problem of labor productivity. . . . Labor productivity decreased 60 percent in coffee production, 44 percent in sugar, . . . and 60 percent in cotton. . . . It is unacceptable that our workers put in two or three hours of work with low work standards, and then earn their whole eight-hour daily salary; that they do half the work and earn twice as much. What economy can withstand that?
—Ministry of Agriculture and Land Reform, 1985

With one week's salary a worker in my company can afford to buy one Coca-Cola. If I double his salary, he can afford two Coca-Colas. Why should he work?
—Manager, state-owned farm, 1985

Between 1937 and 1979 Nicaragua's political and economic landscape had been dominated by the Somoza family. When the regime was finally overthrown by the 1979 revolution, there was a clear perception on the part of many Nicaraguans that a sad chapter in their national history had come to an end. The price that the Nicaraguan people had to pay to bring about this outcome was enormous. In human lives alone, the loss caused by the revolutionary struggle was estimated to be 40,000 in a country of only 2.7 million inhabitants.[1]

Despite all the suffering these losses implied, when a new page

---

1. The U.S. government area handbook for Nicaragua, for example, states: "Human and material losses from the conflict were staggering. An estimated 30,000 to 50,000

Table 20. Economically active population in the farm sector

1978		1984	
Category	No. of people	Category	No. of people
1 Large landowners	2,900	1 Large landowners	N.A.
2 Medium landowners	38,400	2 Medium landowners	N.A.
3 Small landowners	97,300	3 Small landowners	N.A.
4 Poor farmers	157,600	4 Poor farmers:	N.A.
		4a Individual land reform recipients	1,009
		4b Beneficiaries of legalized land titles	33,145
5 Landless permanent workers	60,900	5 Landless permanent workers:	N.A.
		5a State farms	36,000
6 Landless seasonal workers	75,200	6 Landless seasonal workers	N.A.
		7 Sandinista cooperatives	30,000
		Total of available data	100,154
Total	432,300		412,000

Sources: 1978 data is from CIERA/PAN/CIDA, *Informe final del projecto estrategia alimentaria*, vol. 1 (Managua: CIERA, 1984), p. 25. 1984 data is from MIDINRA. Total is from Ministry of Labor (MITRAB).

was eventually turned in July 1979, the general mood was one of optimism and high expectations, which was particularly true for landless peasants and poor farmers. The postrevolutionary government of Nicaragua estimated that in 1978 these poorer segments of the rural population amounted to about 294,000 people. This segment, in turn, was subdivided into poor farmers owning less than seventeen acres of land, landless permanent workers, and landless seasonal workers (see Table 20). Over the years of revolutionary struggle, these peasants had been motivated by the Sandinista party (FSLN) and other opposition groups to join the fight against Somoza with the promise of land ownership and a bright future. FSLN revolutionary propaganda aimed at the peasantry in the 1960s and 1970s had stressed the unnecessary poverty of most Nicaraguans.[2]

Nicaraguans had been killed during the civil conflict . . . and some 100,000 had been injured" (James D. Rudolph, ed., *Nicaragua: A Country Study* [Washington, D.C.: Government Printing Office, 1982], pp. 57–58).

2. Forrest D. Colburn, *Post-Revolutionary Nicaragua: State, Class, and the Dilemmas of Agrarian Policy* (Berkeley: University of California Press, 1986), p. 108. See also the works of the founder and major theoretician of the Sandinista party, Carlos Fonseca, in *Obras*, vols. 1, 2 (Managua: Editorial Nueva Nicaragua, 1981).

Table 21. Estimated distribution of land ownership (%)

Sector	1978	1981	1982	1984	1985
Individual					
Large (over 200 mz[a])	51	31	27	23	23
Medium (50–200 mz)	30	30	30	30	30
Small (10–50 mz)	16	7	7	7	7
Poor (less than 10 mz)	2	1	1	1	2
Cooperative					
Independent production	1[b]	—	—	—	—
Sandinista Agricultural (CAS)	—	1	2	9	9
Credit and Services (CCS)	—	10	10	10	10
State farm (APP)[b]	—	20	23	20	19

Source: MIDINRA.
Note: Estimates vary between different government sources.
[a]One manzana (mz) = 1.7 acres.
[b]Twenty-two cooperatives with an estimated 5,000 members.

As Forrest D. Colburn explains:

> The struggle to overthrow Somoza and the triumph of the revolution deepened a sense of deprivation among peasants and created a sense of hope for a better future. These sentiments were widespread and not limited to those who participated in the struggle to oust the dictator or those organized into the Sandinista rural organization, the Association of Rural Workers (ATC). The FSLN had labored for years to convince peasants that they were being exploited and that a better future awaited them upon the triumph of the revolution. One of the popular FSLN slogans was "land for peasants." Many Nicaraguans thought that after the revolution they would suddenly have everything they never had before, and they would no longer have to work.[3]

In the first four years after the revolution, however, state policy did not encourage land redistribution to individual peasants.[4] Instead, it stressed state ownership of the land and later the development of a cooperative movement under the control of the Sandinista party. Government data, presented in Table 21, indicates that in 1981, state-owned farms, which had not existed in 1978, controlled

---

3. Colburn, *Post-Revolutionary Nicaragua*, p. 107.
4. Michael Zalkin, "Peasant Response to State Grain Policy in Post-Revolutionary Nicaragua: 1979–1984" (Ph.D. diss., University of Massachusetts, Amherst, 1986), p. 140. See also Michael Zalkin, "Food Policy and Class Transformation in Revolutionary Nicaragua, 1979–86," *World Development* 15, no. 7 (1987): 961–984; David Kaimowitz and David Stanfield, "The Organization of Production Units in the Nicaraguan Agrarian Reform," *Inter American Economic Affairs* 39, no. 1 (1985): 51–77; and Marvin Ortega, "Workers' Participation in the Management of the Agro-Enterprises of the APP," *Latin American Perspectives* 12, no. 2 (1985): 69–81.

20 percent of the country's arable land and employed about 36,000 workers. Soon afterwards Sandinista production cooperatives were created, and by 1984 these had about 30,000 members, recruited mostly from among small and poor farmers who already owned some land. These units represented 9 percent of the arable land.

The net result was a large gap between the 136,000 who had been landless workers in 1978 and whose status was unchanged and the 66,000 people, formerly landless peasants, poor farmers, and small landowners, who had benefited from their membership in state farms or Sandinista cooperatives. The people on the disadvantaged side of this gap, at least 70,000 of which were among the poorest of the rural population, aspired to be farmers, not laborers. They felt betrayed by a revolution that had promised them land ownership and had then reneged on its promise. It was mainly from among these deeply disappointed peasants, outside the state and cooperative sector, that the counterrevolutionary movement later recruited its members.

The government often claimed that it had given individual titles of land to 34,154 people between 1979 and 1984 (Table 20, line 4a plus line 4b). These statistics, however, were misleading because 97 percent of these people had not been a part of the landless peasantry in 1978. They were, instead, poor farmers already in possession of land in 1978; all they were given after 1979 was a legal title on land they felt had already been theirs for years. In the first six years following the revolution, only 1,009 new individuals actually received land, and this group obtained less than 1 percent of the arable land.[5]

Many poor and landless peasants found themselves working as wage laborers in Somoza's expropriated estates. These lands had been converted by the postrevolutionary government into state-owned enterprises and were officially designated as the Area of People's Property (APP). Many poor peasants interpreted this designation to mean that those properties were theirs because they had been told for years that they were the people and that the revolution was being fought for their benefit. Consequently, many of these peasants felt, particularly as they had been denied individual ownership of land, that at the very least they could affirm their new "liberated" status by working on state property as they pleased. The drop in labor productivity that ensued from this attitude came to be termed *la vacación histórica,* or the historical vacation. After having

---

5. This point is also made by Nola Reinhardt, "Agro Exports and the Peasantry in the Agrarian Reforms of El Salvador and Nicaragua," *World Development* 15, no. 7 (1987): 950.

been told for so long that they were exploited, wage laborers felt that victory against Somoza had to mean land ownership or, failing that, at least the freedom to take a good vacation.[6] During the year after the revolution, there had been several instances in which poor peasants and landless workers had invaded state-owned farms and large private properties and had begun to create self-managed production units,[7] but the government quickly dissolved them.[8] Nevertheless, until 1985 the vast majority of the peasants assigned to state-owned farms, despite their disappointment, were willing to give *Sandinismo* a chance. For a while, they were willing to experiment with working as wage laborers. They did not, however, put in an eight-hour workday in the fields, as in the past. Instead, they worked about half that amount of time and then returned home, receiving payment for the full eight-hour workday.

Sympathetic as they may have been toward the workers, the new revolutionary leaders understood that this practice could not go on forever. The FSLN, once in power, quickly shifted its emphasis from promoting labor militancy to stressing labor discipline.[9] In a seminal paper written a few months after the revolution, Henry Ruiz, minister of planning and a member of the nine-man Sandinista directorate, addressed this issue in unequivocal terms: "The participation of workers in state-owned enterprises does not mean the breakdown of labor discipline and a failure to perform according to labor standards. It means instead the creation of a new labor discipline where political consciousness is well developed and production is highly efficient."[10]

Despite this and other government statements, the drop in labor productivity immediately following the revolution was not redressed. On the contrary, after 1980 the amount of work performed

---

6. Collins states that "the new order gave [agricultural laborers] an immediate chance for only one tangible good—less work" (Joseph Collins, *What Difference Could a Revolution Make? Food and Farming in the New Nicaragua* [San Francisco: Institute for Food Development and Policy, 1985], p. 75).

7. See Zalkin, "Peasant Response to State Grain Policy," pp. 94–315. See also Kaimowitz and Stanfield, "Nicaraguan Agrarian Reform," pp. 51–77; Carmen Diana Deere, Peter Marchetti, S.J., and Nola Reinhardt, "The Peasantry and the Development of Sandinista Agrarian Policy, 1979–1984," *Latin American Research Review* 20, no. 3 (1985): 75–110; and Ortega, "Management of the Agro-Enterprises of the APP," pp. 69–81.

8. See, for example, Zalkin, "Peasant Response to State Grain Policy," pp. 144–151; and Ortega, "Management of the Agro-Enterprises of the APP," pp. 71–72.

9. Colburn, *Post-Revolutionary Nicaragua*, p. 108.

10. Henry Ruiz, *El papel político del APP en la nueva economía sandinista* (Managua: FSLN, 1980), p. 22.

in state-owned enterprises decreased even further. In numerous instances the average workday dropped to two hours per day, prompting MIDINRA, in its 1985 annual report, to state: "It is unacceptable that our workers [employed by state-owned agribusiness enterprises] put in two or three hours of work with low work standards, and then earn their whole eight-hour daily salary; that they do half the work and earn twice as much. What economy can withstand that?"[11] The attitude of the rural labor force was thought to be partly responsible for decreases in farm output and for increases in food shortages and inflation.[12]

This chapter explores the possibility that the government's "social-wage" policy, whereby increases in nominal wages were discouraged in favor of subsidized food staples and improved social services, effectively removed the managerial instrument of monetary incentives from the hands of state administrators. Over time, this policy led to a substantial fall in the purchasing power of wages and undermined worker motivation, perpetuating the considerable loss in labor productivity that had followed the revolution and the nationalization of Somoza's agribusiness operations. Consequently, correcting state enterprise inefficiencies became an even more arduous managerial task.

Labor Productivity in State Enterprises: Goal and Performance

The postrevolutionary government of Nicaragua had always considered high labor productivity in state-owned agribusiness enterprises to be a vital component of the so-called New Model of Accumulation and the New Economy.[13] These two expressions were

---

11. MIDINRA, *Plan de trabajo, balance y perspectivas, 1985* (Managua: MIDINRA, 1985), p. 35.

12. See, for example, SPP, *Plan económico 1985* (Managua: SPP, 1985), pp. 1–3.

13. Research sources regarding the question of labor productivity in Nicaraguan state-owned agribusiness enterprises after 1979 can be divided into three groups: unpublished government documents, published government documents, and published independent studies. Unpublished government documents represent by far the most important source of information. The following are the most important: MIDINRA, *La recuperación de la jornada laboral en el campo* (Managua: MIDINRA, 1986). MIDINRA, *Documento de estudio sobre la productividad del trabajo* (Managua: MIDINRA, 1980). MIDINRA, *La productividad del trabajo en el sector agropecuario, 1985–1986* (Managua: MIDINRA, 1986). MIDINRA, *Resultados de la productividad laboral en la labor de cosecha de los cultivos de agroexportación: Ciclo productivo 1985/1986* (Managua: MIDINRA, 1986). MIDINRA, *Informe-diagnóstico sobre la productividad en la región de Matagalpa* (Managua: MIDINRA, 1980). MIDINRA, *Informe sobre las normas de café a pagar en*

las cosechas 1980-81. Fuerza laboral y producción estimada (Managua: MIDINRA, 1980). MIDINRA, *Informe sobre las normas a pagar en la cosecha algodonera 1980-81: Fuerza laboral y producción estimada* (Managua: MIDINRA, 1980). MIDINRA, *Propuesta de ajuste a las normas de trabajo manuales del cultivo: Caña de azúcar* (Managua: MIDINRA, 1986). MIDINRA, *Segundo análisis sobre el comportamiento de la normativa laboral, jornada de trabajo y salario por rubro* (Managua: MIDINRA, 1986). MIDINRA, *Cuarto análisis sobre el comportamiento de la normativa laboral, jornada de trabajo y salario por rubro* (Managua: MIDINRA, 1986). MIDINRA, *Normas de trabajo para labores manuales: Ciclo agrícola 1986-1987* (Managua: MIDINRA, 1986). MIDINRA, *Normativa laboral de cosecha 1986/88 de los cultivos de agroexportación* (Managua: MIDINRA, 1986). AGROMAQ, *Evaluación de los indicadores económicos de la empresa* (Managua: MIDINRA, 1987). Ministry of Labor, *Propuesta de normativa salarial para dirigentes de empresa* (Managua: MITRAB, 1984). MIDINRA, *Política salarial para el personal de dirección* (Managua: MIDINRA, 1984). MIDINRA, *Propuesta salarial para directores de empresas, responsables de complejo, responsable de UPE* (Managua: MIDINRA, 1984). DGATM, *Incentivos periódicos y finales* (Managua: MIDINRA: 1987). MIDINRA, *Planes de incentivo laboral: ENIA, AGROINRA, ENFARA, TIMAL, DGIFA* (Managua: MIDINRA, 1987). MIDINRA, *Situación de los salarios mínimos* (Managua, MIDINRA, 1983). MIDINRA, *Breve análisis salarial, 1981-1982: Nivel institucional* (Managua: MIDINRA, 1983). MIDINRA, *Apuntes sobre la problemática y política salarial: 1983: Síntesis de la política salarial MITRAB-MIPLAN* (Managua: MIDINRA, 1983). MIDINRA, *Estudio por rubro e impacto económico de producirse un aumento salarial* (Managua: MIDINRA, 1983). MIDINRA, *Testimonios de proyectos de incentivos implementados en el sector agropecuario durante los primeros años de la revolución* (Managua: MIDINRA, 1984). MIDINRA, *Propuesta de nuevo sistema de remuneración salarial en el sector agropecuario* (Managua: MIDINRA, 1984). MIDINRA, *Análisis del comportamiento salarial en el agro* (Managua: MIDINRA, 1981). MIDINRA, *Problemática de las empresas del sector agropecuario adscritas al MIDINRA* (Managua: MIDINRA, 1985). Central Sandinista de Trabajadores, *Análisis general sobre producción, productividad, salario, abastecimiento, estado político organizativo de los trabajadores e impacto del bloqueo económico* (Managua: CST, 1985). CST, *Trabajo para impulsar la normación* (Managua: CST, 1985). Ministry of Planning, *Política salarial* (Managua: MIPLAN, 1980). CST, *El sistema de organización del trabajo y los salarios (SNOTS)* (Managua: CST, 1985). CST, *Productividad y salario: Análisis del avance del cumplimiento de los acuerdos del segundo balance* (Managua: CST, 1985). CIERA, *Principales problemas que están incidiendo sobre la producción agropecuaria* (Managua, MIDINRA, 1985). CST, *Conclusiones sobre la relación salario-productividad-abastecimiento* (Managua: CST, 1985). MIDINRA, *Principales causas estructurales que afectan negativamente la eficiencia y la situación económica-financiera de las empresas adscritas al MIDINRA* (Managua: MIDINRA, 1985). MIDINRA, *Experiencia piloto de gestión en las empresas de reforma agraria* (Managua: Consejo Nacional de Reforma Agraria, 1983).

The next series of citations are among those published government documents related to labor productivity in state-owned enterprises: Henry Ruiz, *El papel político del APP.* MIDINRA, *Plan de trabajo, balance y perspectivas,* p. 35. INEC, *Indicadores del sistema de trabajo y salario* (Managua: INEC, 1983). MITRAB, *Algunas consideraciones teóricas para la interpretación de la problemática salarial en Nicaragua* (Managua: MITRAB, 1980). MITRAB, *Guía de derechos laborales* (Managua: MITRAB, 1979). MITRAB, *La convención colectiva en Nicaragua* (Managua: MITRAB, 1984). MITRAB, *Boletín de estadística sociolaboral* (Managua: MITRAB, 1981). FSLN, *Habla la dirección de la vanguardia* (Managua: FSLN, 1981). MITRAB, *Balances de fuerza de trabajo* (Managua: MITRAB, 1983). SPP, *Empleo e ingreso rural y agropecuario* (Managua: SPP, 1983). Comisión Nacional sobre Migración Campo-Ciudad, "Problemas y perspectivas de la migración campo-ciudad," *Revolución y desarrollo* 3 (July–September 1985): 19–23. Ricardo Coronel Kautz, "La normación del trabajo y salario y la difícil situación de nuestras empresas estatales," *Revolución y desarrollo* 2 (July–September 1984): 25–26. FSLN,

widely used by government policymakers to designate the overall strategy of assigning to the state the main role in the generation of

---

"Las doce tareas del pueblo nicaragüense," *Pensamiento propio* 22 (April 1985): 44. Secretariat of Propaganda and Political Education of the FSLN, *La propaganda de la producción* (Managua: 1980). Victor Tirado, "Improvement in the Situation of Workers Is the Task of the Workers Themselves," in *Nicaragua: The Sandinista People's Revolution*, ed. Bruce Marcus (New York: Pathfinder Press, 1985). Jaime Wheelock Román, *Entre la crisis y la agresión: La reforma agraria sandinista* (Managua: MIDINRA, 1984), pp. 109–113. Jaime Wheelock Román, "El movimiento sandinista y la lucha de clases," *Encuentro* 15 (1980): 21–38. Jaime Wheelock Román, *Discurso en reunión sostenida con trabajadores y directores del sector arrocero* (Managua: Cementera, May 4, 1985). Jaime Wheelock Román, *Exposición sobre la gestión de los trabajadores* (Managua: Auditorio del INSSBI, May 3, 1983). Jaime Wheelock Román, *Asamblea Nacional de los Trabajadores del Algodón* (Managua: Complejo Cívico Camilo Ortega, June 9, 1985). Jaime Wheelock Román, "The Sandinista Front Is the Organization of the Working People," in *Nicaragua: The Sandinista People's Revolution*, ed. Marcuso.

Finally, the remaining citations are among those published independent studies on labor productivity in postrevolutionary Nicaragua. Richard Stahler-Scholk, *Empleo, salarios y productividad en la revolución popular sandinista* (Managua: CRIES, 1986). Richard Stahler-Scholk, *La normación del trabajo en Nicaragua, 1983–1986* (Managua: ANICS, 1986). Richard Stahler-Scholk, *Política salarial en Nicaragua, 1979–1985* (Managua: CRIES, 1985). Kaimowitz and Stanfield, "Nicaraguan Agrarian Reform," pp. 51–77. INIES, "El concepto de productividad," *Cuadernos de investigación*, no. 1 (1986). INIES, "La producción y la productividad," *Foro-debate*, no. 1 (1987). Deere, Marchetti, and Reinhardt, "Sandinista Agrarian Policy," pp. 75–110. Ortega, "Management of the Agro-Enterprises of the APP," pp. 69–81. Christian Guillén, *Comentarios sobre la productividad y los salarios* (Managua: INIES, 1981). A. Martínez, *Situación económica global y empleo* (Managua: PREALC, 1982). INIES, "No basta conciencia política para elevar productividad en el campo," *Boletín socio-económico*, no. 3 (May–June 1987): 3. Arts Diego and Teyo van de Schoot, *La organización del trabajo y participación obrera en una finca cafetalera estatal* (Managua: UNAN, Departamento de Economía Agrícola, 1985); a case study of Agricultural Enterprise, UPE Sta. Martha, Chale Haslam. Carlos Bendaña, "Reflexiones sobre la participación popular," *Pensamiento propio* 15 (July 1984): 32–35. Forrest D. Colburn, *Post-revolutionary Nicaragua*, pp. 103–120. Collins, *What Difference Could a Revolution Make?*, pp. 69–78. Carmen Diana Deere and Peter Marchetti, S.J., "The Worker-Peasant Alliance in the First Year of the Nicaraguan Agrarian Reform," *Latin American Perspectives* 8 (Spring 1981): 40–73. Silvio de Franco, "Strategies of Subsistence and Interdependency: Their Implications for Agrarian Development Policy in Nicaragua," Working Paper (Managua: INCAE, 1982). Laura J. Enríquez, "The Dilemmas of Agroexport Planning," in *Nicaragua: The First Five Years*, ed. Thomas W. Walker (New York: Praeger, 1985). Richard Harris, "A Commentary on the Contemporary Conjuncture in Nicaragua: Response to James Petras," *Latin American Perspectives* 10 (Winter 1983): 114–116. Lucio Jiménez, "Defense Can Be Assured Only by Increasing Production," in *Nicaragua: The Sandinista People's Revolution*, ed. Marcus. Marvin Ortega, "La participación obrera en la gestión de las empresas agropecuarias del APP," in *La Revolución en Nicaragua*, ed. Richard Harris and Carlos M. Vilas (Mexico City: Era, 1985). Marifeli Pérez-Stable, "The Working Class in the Nicaraguan Revolution," in *Nicaragua in Revolution*, ed. Thomas W. Walker (New York: Praeger, 1982). James Petras, "Workers' Democracy: The Key to Defending the Revolution and Developing the Productive Forces. Response to Richard Harris," *Latin American Perspectives* 10 (Winter 1983): 117–119. Third National Assembly of Unions, "Economic Production Is the Rear Guard of the Battle Front," in *Nicaragua: The Sandinista People's Revolution*, ed. Marcus. Zalkin, "Food Policy and Class Transformation," pp. 961–984.

economic surplus.[14] According to government policymakers, any protracted productivity failure in the state sector would inevitably have serious economic implications for the advancement of the New Economy, the consolidation of the state as the engine of economic growth, and the feasibility of the government's ambitious program of social and economic restructuring.[15]

Although labor productivity problems during an initial organizational phase were considered unavoidable, the sheer magnitude of the drop in the length of the average workday among rural laborers immediately following the successful outcome of the revolution caught the government by surprise.[16] In 1981, Carmen Diana Deere perhaps best summarized what was at stake:

> Preliminary indications in the summer of 1980 showed that the costs of the state farms were running far above what was expected. The formidable task of organizing the state sector with limited numbers of trained personnel was certainly one reason; there were also reports of a nearly 30% decrease in the length of the working day on the state farms. The state sector can hardly afford such an [implicit] increase in the wage. A reduced social surplus not only limits the state's ability to deliver on promises of increased social infrastructure; but should the state become a drag on the national economy, the model of accumulation, itself, would be in jeopardy.[17]

The reduction in the length of the workday mentioned by Deere had actually been closer to 50 percent, as the average number of hours of work performed in a day decreased from eight to about four.[18] A number of MIDINRA internal reports attempting to explain what had happened were unanimous in their observations:

> With the triumph of the revolution, the basis upon which labor discipline rested in the countryside was violently shaken . . . contributing to a drastic rupture of labor discipline in agriculture. . . . As a result, the nation experi-

---

14. See Ruiz, *El papel político del APP;* and George Irvin, *Nicaragua: Establishing the State as the Centre of Accumulation* (The Hague: Institute of Social Studies, 1982).
15. Ruiz, *El papel político del APP;* see also Secretariat of Propaganda and Political Education of the FSLN, *Propaganda de la producción;* FSLN, *Habla la dirección de la vanguardia;* and CST, *Sobre los problemas económicos y las tareas del sindicato* (Managua: FSLN, 1984).
16. See, for example, MIDINRA, *Documento de estudio sobre la productividad del trabajo;* MIDINRA, *La recuperación de la jornada laboral en el campo;* and MIDINRA, *La productividad del trabajo en el sector agropecuario.*
17. Carmen Diana Deere, "Nicaraguan Agricultural Policy: 1979–1981," *Cambridge Journal of Economics* 5 (1981): 199.
18. MIDINRA, *Documento de estudio sobre la productividad del trabajo,* p. i and pt. 5, p. 6.

enced a general decline in labor productivity, which expressed itself in a very sharp manner in the countryside in the form of a drastic reduction in the number of hours of actual work performed [by a farm laborer] in a day.[19]

Although the decrease in labor productivity that immediately followed the revolution was a generalized phenomenon across the nation, government documents indicated that the largest drop in labor productivity had occurred among state-owned agribusiness enterprises.[20] Among these enterprises, labor productivity had decreased 60 percent in the case of coffee, 44 percent in the case of sugarcane, and 60 percent in the case of cotton.[21]

This problem was all the more disconcerting given that one of the stated purposes for creating the APP had been precisely the prevention of a drop in labor productivity that the government believed would otherwise have followed had confiscated properties been parceled.[22] Indeed, the new government considered productivity, along with the goals of reviving production as rapidly as possible, creating more jobs, continuing to produce export crops, and—most important of all—establishing the state as the center of capital and profit accumulation, so critical as to justify the political risk of not distributing expropriated land to poor and landless peasants.[23]

If it had been momentarily taken by surprise, the government lost no time in recovering the initiative. Fending off opponents in 1980 on both the left and the right, the FSLN ceased to stress the unnecessary poverty of the majority of the population and began to explain the need for austerity and production.[24] Propaganda manuals distributed by the FSLN to the leaders of its mass organizations switch-

19. MIDINRA, *La recuperación de la jornada laboral en el campo*, pp. 2, 12. See also MIDINRA, *Documento de estudio sobre la productividad del trabajo*; MIDINRA, *La productividad del trabajo en el sector agropecuario*; MIDINRA, *Labor de cosecha de los cultivos de agroexportación*; and MIDINRA, *Productividad en la región de Matagalpa*.

20. See MIDINRA, *Documento de estudio sobre el problema de la productividad del trabajo* (Managua: MIDINRA, 1980). The report stated: "We are aware of the enormous problem represented in these moments by the decrease in labor productivity that is occurring in various sectors of our economy, primarily in the APP" (pt. 4, p. 1).

21. MIDINRA, *Plan de trabajo, balance, y perspectivas*, p. 35.

22. See, for example, James E. Austin, Jonathan Fox, and Walter Krüger, "The Role of the Revolutionary State in the Nicaraguan Food System," *World Development* (January 1985), p. 19; Collins, *What Difference Could a Revolution Make?*, pp. 60–63; Zalkin, "Food Policy and Class Transformation," pp. 963–964; and Deere, Marchetti, and Reinhardt, "Sandinista Agrarian Policy," pp. 79–80.

23. Collins, *What Difference Could a Revolution Make?*; Zalkin, "Food Policy and Class Transformation"; and Deere, Marchetti, and Reinhardt, "Sandinista Agrarian Policy."

24. See, for example, Colburn, *Post-Revolutionary Nicaragua*, p. 108; and Collins, *What Difference Could a Revolution Make?*, p. 71.

ed from promoting labor militancy to emphasizing labor discipline.[25]

Particular effort was placed on educating the labor force in the state sector. Government leaders explained to APP workers that state ownership of the means of production signified that profits would no longer benefit a small minority of wealthy landowners but would accrue to the entire population, particularly those segments in greater need. Workers in state farms were not to be considered wage laborers. "They are producers of social wealth," explained MIDINRA minister Jaime Wheelock, "and the consciousness of the producer is quite different from that of the wage laborer. . . . He knows that each stroke of the machete is no longer to create profits for a boss, but perhaps to create a new pair of shoes for a barefoot child who may be his own."[26] The performance of this producer of social wealth was shown to be critical. Without high worker productivity, state-owned enterprises could not generate the surplus needed to make new investments, create additional employment, and increase a social wage made up of accessible health services, education, housing, and transportation.[27]

Subsequently, activists of the FSLN party and leaders of the Sandinista labor union, Central Sandinista de Trabajadores (CST), admitted that they had at times misunderstood their role in the production sector. As military activities became more pronounced in 1983, this error consisted of a failure to see the economy as the rear guard of the Sandinista army in its fight against the counterrevolution. As a result, the FSLN leadership criticized activists and union leaders for promoting labor productivity only through such limited gestures as volunteering one day of work per week without pay, when at the same time these activists and union leaders had failed to show effective control over labor productivity, labor discipline, and the efficient use of resources on the other regular workdays. Party activists and union leaders were also faulted "for having gone to the extreme of paralyzing production for the sake of obtaining higher wages and improved food supplies."[28]

25. Colburn, *Post-Revolutionary Nicaragua*, p. 108; see also Ruiz, *El papel político del APP*; Secretariat of Propaganda, *Propaganda de la producción*; FSLN, *Habla la dirección de la vanguardia*; and CST, *Sobre los problemas económicos y las tareas del sindicato*.
26. Jaime Wheelock Román, as quoted by Collins, *What Difference Could a Revolution Make?*, p. 75.
27. Ruiz, *El papel político del APP*; and Secretariat of Propaganda, *Propaganda de la producción*, p. 7.
28. CST, *Sobre los problemas económicos y las tareas del sindicato*, pp. 9–10.

There was a consensus among the various government reports that despite these and other efforts,[29] labor productivity had continued to fall between 1980 and 1985.[30] An early MIDINRA survey of state-owned agribusiness enterprises reported that "during the implementation of the 1980 Economic Plan, we have observed a marked drop in labor productivity [in the APP]. . . . *The fact that during a day workers work only half of the hours they are paid for . . . has very important negative consequences for the national economy, since on the one hand it affects the total value of production, and on the other it affects the generation of profits and the capacity to accumulate capital.*"[31] Likewise, a subsequent MIDINRA study relating mostly to the APP reported, "Since the triumph of the Sandinista Popular Revolution a gradual reduction in labor productivity and in the number of daily hours worked began in the farm sector, so that by 1983–84 it is estimated that the decrease amounted to 50 percent on average in the various crops."[32] Another MIDINRA report showed graphically how the daily amount of work performed in the countryside went from six hours in early 1979 to four hours one year later and further decreased to about two and one-half hours in 1982.[33] Excerpts of this report formed the basis of a major article on labor productivity in the Sandinista party newspaper, *Barricada*, on June 17, 1986. A more limited survey made in 1985 by CST reported, "Labor productivity decreased 8 percent annually . . . [and] the average number of hours actually worked in a day decreased 6 percent during the past two years."[34] Similarly, Comandante Víctor Tirado stated in a speech to labor union leaders: "Between 1983 and 1984 there was a well-known deterioration of labor productivity. That means that instead of improving our discipline and efficiency, we have worsened them. . . . It is good that workers consider self-critically such numbers as the increase of unjustified

---

29. See, for example, discussions of the effort after 1983 of introducing a new system of mandatory salary scales and regulations, the so-called SNOTS (*Sistema Nacional de Organización del Trabajo y los Salarios*), in three works by Richard Stahler-Scholk, *Empleo, salarios y productividad*; *La normación del trabajo*; and *Política salarial*.

30. See, for example, SPP, *Plan económico 1985*, pp. 1–3.

31. MIDINRA, *Documento de estudio sobre el problema de la productividad del trabajo*, p. i and pt. 5, p. 6, underlined in original.

32. MIDINRA, *Normas de trabajo para labores manuales*, p. 1.

33. MIDINRA, *La recuperación de la jornada laboral en el campo*, p. 12.

34. CST, *Análisis general*, p. 2.

absenteeism to 50 percent, whereas in 1983 it was only 39 percent... in 1983 hours worked represented 72 percent of the work day, but in 1984 they were only 67 percent."[35] Yet another assessment of labor productivity in state-owned agribusiness enterprises concluded: "With the triumph of the revolution in Nicaragua productivity [in the state sector] . . . fell in an alarming way. . . . . Tasks that prior to the revolution were carried out by one man, in the new situation took two or three men. Production costs increased and the hours actually worked in 1984 were only two or three per day. These problems caused serious damages to the economy."[36] Six years after the revolution, the government was still struggling with the problem. The 1985 National Economic Plan stated that one of the reasons why the economy was expected to deteriorate further that year was "a drastic drop in average labor productivity, particularly in the countryside, which results in cost increases and a decrease in production."[37]

The concept of the "historical vacation" was useful in attempting to explain the initial drop in labor productivity among rural workers immediately following the revolution. Complex forces, including the early euphoria for the victory over Somoza, the loss of fear of being brutalized or fired by their employers, and disappointment over the creation of state farms at the expense of giving land entitlements directly to individual peasants, all played a part in explaining the apparent lack of commitment of the workers in state farms. After several years of decreasing labor productivity and the failure of numerous government attempts to redress the situation, however, it was legitimate to ask whether some other element was at play. To answer this question, I now turn to an analysis of wage rates and their impact on labor productivity in the context of a state farm operating in the rice commodity system.

---

35. *Barricada*, May 16, 1985.
36. INIES, "Conciencia política," p. 3.
37. SPP, *Plan económico 1985*, p. 3.

Wage Rates and Labor Productivity: The Case of a State Rice Farm[38]

Rice Production: An Important Success in Jeopardy

One of the fundamental objectives pursued by the Sandinista government was to redress the historic imbalance between agroexport and basic grains production. Food security as a policy objective meant, first and foremost, self-sufficiency in rice, corn, and beans, the three grains that make up the basic daily diet of the Nicaraguan worker.[39] The importance of basic grains production did not imply, however, a "food versus export crops" dilemma.[40] A combination of circumstances, including a highly favorable person-to-land ratio, meant that Nicaragua was not immediately forced to choose one sector at the expense of the other.[41] It was important, though, that basic grain production be increased as rapidly as possible in order to reduce the dependence on imports and to supply a rapidly growing population whose demand for food was expected to grow in response to new policies aimed at improving its access to food.[42]

---

38. In addition to the records of the SRF and personal interviews primarily held between 1984 and 1987, the following are some of the sources used in this section regarding rice production: James E. Austin, *Marketing Adjustments to Production Modernization: The Case of the Nicaraguan Rice Industry* (D.B.A. diss., Harvard University, 1972). MIDINRA, *Evolución histórica y perspectiva, 1988–1990: Granos básicos y hortalizas* (Managua: MIDINRA, 1987). MICOIN, *Sistemas de comercialización: Productos básicos de consumo popular,* vol. 2, *Granos básicos, Arroz,* pp. H-1–H-47. MIDINRA, *Programa de arroz, ciclo 1986/87* (Managua: MIDINRA, 1986). MIDINRA, *Programa de producción arrocera 1987/88* (Managua: MIDINRA, 1987). MIDINRA, *Perfil estratégico para la organización de la producción y el intercambio de granos básicos de la IV región* (Granada: MIDINRA, 1987). MIDINRA, *Perfil estratégico para la organización de la producción y el intercambio de arroz en el territorio de Malacatoya* (Granada: MIDINRA, 1986). MIDINRA, *Reunión con productores de arroz: Síntesis de reuniones de trabajo entre el ministro de desarrollo agropecuario y reforma agraria, Jaime Wheelock Román y los productores nacionales* (Managua: MIDINRA, 1985). Alfred H. Saulniers, *State Trading Organizations in Expansion: A Case Study of ENABAS in Nicaragua* (Austin: University of Texas, 1987). BCN, *Granos básicos* (Managua: BCN, 1973). Sonia Agurto Vílchez and Peter Utting, "La comercialización de alimentos básicos en Nicaragua," in *Comercialización interna de los alimentos en América Latina,* ed. CIAT (Cali: CIAT, 1985). COSEP, *El arroz y los frijoles,* Memorandum de la Presidencia No. 10 (Managua: COSEP, 1986). MIDINRA, *Producción, distribución, y consumo de los principales productos de consumo básico* (Managua: MIDINRA, 1983). CIERA, *Distribución y consumo popular de alimentos en Managua* (Managua: MIDINRA, 1983). Programa Alimentario Nacional (PAN): for a bibliography of PAN documentation consulted, see n. 52, Chapter 2.

39. See, for example, MIDINRA, *Marco estratégico del desarrollo agropecuario* (Managua: MIDINRA, 1983).

40. Austin, Fox, and Krüger, "Role of the Revolutionary State," p. 19.

41. Ibid.; and Deere, *Nicaraguan Agricultural Policy,* p. 200.

42. See, for example, James E. Austin and Jonathan Fox, "Food Policy," in *Nicaragua:*

Six years after the revolution it was apparent that the desired growth in basic grains had not occurred and that the cumulative domestic supply of rice, corn, and beans was still below prerevolutionary levels (Table 22). Although considerable discrepancies existed in the production statistics for basic grains of different government institutions—Table 22 reports just three such discrepancies in lines 4, 5, and 6—all sources agreed that basic grains production between the crop years 1979–80 and 1984–85 had never achieved the prerevolutionary level of 1978–79. The possible exception, according to only one government source, was the crop year 1983–84, when it was claimed that production slightly exceeded the 1978–79 level (Table 22, line 5). Moreover, imports of basic grains between 1979 and 1985 had not been reduced (Table 22). Contrary to earlier official indications, by late 1985 it was clear that imports of rice, corn, and beans continued to represent about one-third to one-half of domestic production and were expected to increase in 1986 and 1987, as the gap between domestic supply and demand continued to widen.[43]

This overall disappointing outcome appeared to be related to a number of highly complex interacting phenomena for which only partial explanations have yet been given. Available evidence indicated numerous negative forces at play, among them the counterrevolutionary war,[44] a counterproductive basic grains price policy,[45] adverse weather conditions during the 1982–83 crop year,[46] a shift in

---

The First Five Years, ed. Thomas W. Walker (New York: Praeger, 1985), pp. 399–411.

43. See, for example, MIDINRA, *Granos básicos y hortalizas*. On rice alone, the document noted that "since 1984, in order to balance the [rice production] deficit, the government had to rely on increasing imports and donations [of rice] which amount to 2.4 million hundredweights, valued at US$ 42.2 million" (p. 20).

44. See, for example, E. V. K. Fitzgerald, "An Evaluation of the Economic Costs to Nicaragua of U.S. Aggression: 1980–1984," in *The Political Economy of Revolutionary Nicaragua*, ed. Rose J. Spalding (Boston, Allen and Unwin, 1987), pp. 195–216.

45. The problems with basic grains price policy are perhaps best summarized by Jaime Wheelock Román, minister of agriculture, when he stated: "The day that water is given for free nobody is going to be willing to pay for it; the day that beans are given away for free, nobody is going to be willing to produce them, that is logical. If someone gives to me beans at no charge, and I am a producer of beans, why should I continue to produce beans? I prefer to receive them as gifts" (MIDINRA, *Plan de trabajo, balance y perspectivas*, p. 17). This statement refers to a policy of basic grains consumer subsidies that made retail prices lower than farm-gate prices. This generated a "basic grains recycling" process, whereby grain producers, instead of producing, bought grain from state outlets only to resell it to the same outlets at a profit, until grain shortages and skyrocketing subsidies forced the government in 1984–85 to reconsider the policy.

46. See, for example, United Nations, *Repercusiones de los fenómenos meteorológicos de 1982 sobre el desarrollo económico y social de Nicaragua* (Mexico: CEPAL, 1983), p. 32.

Table 22. Basic grains production, imports, and exports, 1977–1985

	1976–77	1977–78	1978–79	1979–80	1980–81	1981–82	1982–83	1983–84	1984–85
1 Rice production (000 cwt)[a]	825	1,050	1,300	937	1,399	1,995	1,634	1,911	1,691
2 Corn production (000 cwt)[a]	4,371	3,942	5,525	3,750	3,995	4,199	2,378	2,159	2,689
3 Beans production (000 cwt)[a]	1,177	894	1,206	862	724	904	686	594	439
4 Total basic grains production (BCN + INEC)[a]	6,373	5,886	8,031	5,549	6,118	7,098	4,698	4,664	4,819
5 Total basic grains production (BCN + SPP)[b]					6,478	7,302	7,121	8,243	7,274
6 Total basic grains production (BCN + MIDINRA)[c]					5,395	7,046	6,772	7,975	7,792
7 Rice imports[c]	1	3	16	816	732	420	99	708	717
8 Corn imports[c]	460	117	375	1,026	727	534	2,968	382	744
9 Bean imports[c]	2	2	21	267	577	54	36	234	278
10 Total grain imports[c]	463	122	412	2,109	2,036	1,008	3,103	1,324	1,739
11 Total grain exports[c]	181	6	158	4	52	51	36	4	—
12 Net grain imports (10 – 11)	282	116	254	2,105	1,984	957	3,067	1,320	1,739
13 Apparent grain consumption (4 + 12)	6,655	6,002	8,285	7,654	8,102	8,055	7,765	5,984	6,558
14 Estimated population (million)	2.5	2.6	2.7	2.8	2.9	3.0	3.1	3.2	3.3
15 Per capita apparent consumption (13 + 14, lbs)	266	230	306	273	279	268	250	187	198
16 Per capita domestic production (4 ÷ 14, lbs)	254	226	297	198	210	236	151	145	146

Sources: [a]BCN (1976–77 to 1979–80) and INEC (1980–81 to 1984–85); [b]MIPLAN and SPP; [c]MIDINRA.
Note: In this table, basic grains include only rice, corn, and beans.

1983 away from a strategy of small, communal grain production toward state capitalist grain production,[47] deteriorating rural-urban terms of trade,[48] decreasing supplies of manufactures to small rural producers,[49] and a state policy that treated a highly diversified peasantry as a homogeneous unit.[50]

Disaggregated production data revealed one exception to this generally bleak picture. In contrast to the substantial decline in corn and bean production, rice output had shown significant gains since 1979. A number of circumstances made this success particularly important to government food policymakers. First, rice was an almost omnipresent component of Nicaraguan meals, soups, sweets, beverages, and other food products. The effect of a rice shortage on the food habits of the population would be likely to be more pervasive than would that of other grains. Second, most of the increase in rice output came from irrigated fields in the central and western plains using modern mechanized cultivation practices. This fact implied that the government could rely on a growing source of grain supply that was relatively less vulnerable to adverse weather conditions and to the effects of the counterrevolutionary war than was the traditional nonmechanized upland rice, corn, and bean cultivation of the agricultural frontier and the interior (Figure 1). Third, state-owned farms produced the bulk of the irrigated rice, and most rice procurement, processing, and distribution was controlled by other state agribusiness enterprises (Figure 5). This situation allowed the state to have greater control over the rice commodity system than it had over other basic grains, facilitating the implementation of a number of government food programs. Fourth, the successful increase in rice output was even more valuable given the persistent problems in corn and bean production and the acute shortage of foreign exchange for importation of grains. If an increase in basic grains imports was to be averted, it was imperative to maintain, if not increase, the gains in rice production.

Government production data indicated, however, that domestic

---

47. See, for example, Zalkin, "Peasant Response to State Grain Policy."

48. See, for example, Zalkin, "Peasant Response to State Grain Policy"; and UNRISD, "Urbanization and Food Systems Development in Nicaragua," in *Food Systems and Society: Problems of Food Security in Selected Development Countries* (Geneva: United Nations, 1986), pp. 190–201.

49. See, for example, MIDINRA, *Plan de trabajo, balance y perspectivas*, p. 17; and Zalkin, "Peasant Response to State Grain Policy."

50. See, for example, Zalkin, "Peasant Response to State Grain Policy"; and UNRISD, "Urbanization and Food Systems Development," pp. 190–201.

Figure 5. The structure of the Nicaraguan rice commodity system: estimated state share of output, 1983

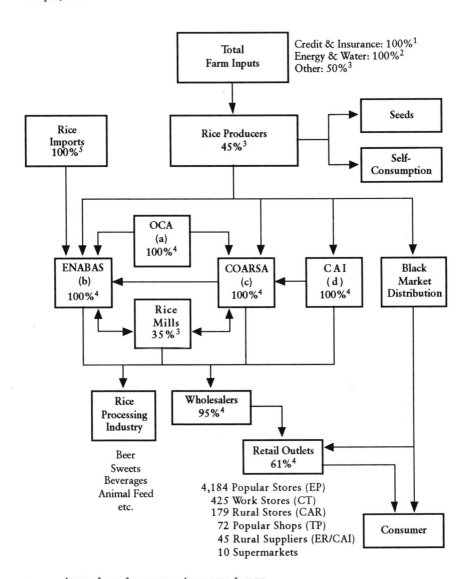

Sources: [1]BCN; [2]INE; [3]MIDINRA; [4]MICOIN; [5]MICE.
(a) State enterprise regulating rice distribution (Oficina Comercializadora de Arroz).
(b) State enterprise procuring, importing, processing, and distributing basic grains.
(c) Private national rice distributor operating under OCA state control. MICOIN considers its distribution part of the state network (Comercializadora de Arroz S.A.).
(d) State rural wholesalers (Centros de Abastecimiento Intermedio, CAI).

rice output peaked in the crop year 1983–84, and began to fall thereafter (Table 22). In the crop year 1984–85, rice production decreased by about 220,000 hundredweights.[51] This fall in rice production came mostly from irrigated rice fields.[52] Internal government reports attributed this decrease in output to poor maintenance of irrigated fields; a lack of care in the preservation of canals, irrigation dams, irrigation equipment, and water pumps; and poor organization of the services supporting rice production.[53]

Field research conducted on one of the model state-owned rice farms responsible for previous output increases attempted to identify what impact, if any, wage rates had had on these production problems.

Wage Policy: Cash Wage versus Social Wage

In 1979 the government conceived the idea of a "social wage," along with the policy built around it, on the basis of two fundamental considerations: that the economy had been disrupted by the revolution and that the commitment to improve the conditions of the poorer segments of the population must be honored. In its essential elements, the concept was based on the premise that the government's most important immediate goal was to "reactivate the economy for the benefit of the people." This end was to be achieved, first, by improving the conditions of the poorest segments of society through an increase in employment and, second, by a more equitable redistribution of goods and services, primarily through improved social services.[54]

It followed that, insofar as possible, nominal cash wages would not be increased. Aside from a strict enforcement of the minimum wage law, any incremental benefit to the poorer segments of society would come from either additional employment or an increase in social benefits, namely, subsidized food, health, education, housing, and transportation.[55] The resulting social wage was therefore conceived as the sum of two components: a fixed component represent-

---

51. See, for example, MIDINRA, *Programa de arroz*, and MIDINRA, *Granos básicos y hortalizas*.
52. MIDINRA, *Programa de arroz*.
53. Ibid., p. 13.
54. Ministry of Planning, *Programa de reactivación económica en beneficio del pueblo, 1980* (Managua: MIPLAN), p. 98.
55. Ministry of Planning, *Programa de reactivación económica*.

ed by the nominal cash wage and a growing component consisting of social benefits, including an increasing access to improved public services.

In the farm sector, the cash component was calculated on the basis of the minimum hourly wage, which was set at 2.65 cordobas in July 1979 and left unchanged until May 1981. The daily wage, therefore, amounted to 21.20 cordobas, and the basic monthly salary in the countryside was equal to 644.43 cordobas. To this basic salary were added social benefits (*prestaciones sociales*), which consisted of one additional daily salary on Sundays (the *séptimo*, which resulted in a double daily pay once a week) and pro rata payments for vacations (*doceavo*) and for the year's end additional monthly salary (*treceavo*). Most workers received in addition a *permuta*, which consisted of a daily food, housing, and transportation allowance of 13.20 cordobas. When social compensations and living allowances were added to the basic monthly salary, a worker earning the minimum hourly wage of 2.65 cordobas received a total monthly compensation of 1,237.63 cordobas. Between June 1981 and May 1983, this total minimum monthly compensation was raised by 4.6 percent, to 1,399.70 cordobas. After 1983, it became subject to a new system of mandatory salary scales and regulations, the so-called National Organizing System of Labor and Wages, or SNOTS (*Sistema Nacional de Organización del Trabajo y los Salarios*), and was increased a number of times.[56]

Nominal Wages versus Real Wages: A Dilemma of the State Rice Farm

In the case of the State Rice Farm (SRF), complete payroll records were available only beginning with the year 1982. This situation was common among state-owned enterprises, as most of them had been created between late 1979 and 1980 under very difficult administrative circumstances. Understandably, administrative priorities during that period had been placed on getting production back to normal. "In 1980," explained one SRF manager, "we were so short of administrative personnel that we had to make choices. It was better for us to plant one acre of rice without keeping a record of it than to write on a piece of paper that no acre was planted."[57]

---

56. See, for example, the works of Stahler-Scholk, *Empleo, salarios y productividad; La normación del trabajo;* and *Política salarial.*

57. Interview, June 1985.

SRF was organized like a typical state-owned farm (see Figure 6). The enterprise had over fifty thousand acres of land divided into five production complexes (*complejos*), each made of five or more smaller farms called state production units, or UPEs (*unidades de producción estatal*).[58] Each UPE had its own administrator, who was responsible for its day-to-day operations and reported to the manager of the complex who, in turn, reported to the enterprise's director. UPEs growing the same crop were usually grouped under the same complex, so as to realize, whenever practical, a certain amount of specialization among complexes. Seven other departments coordinating finances, personnel, supplies, machinery, a support staff of basic grain agronomists, and two processing plants completed the company's organization. The enterprise's director was accountable to the MIDINRA regional director of the area where the main activities of the enterprise were located.

The enterprise's main lines of activity were the production of rice and other basic grains, as well as cattle ranching. The company also operated two plants, one for rice milling and the other for processing animal feed. When SRF was created toward the end of 1979, it was endowed with some of the best irrigated rice fields in Nicaragua. These fields had been expropriated by MIDINRA when their owners fled the country shortly after Somoza's fall. Since then, the irrigated rice fields had been expanded into an area owned by SRF contiguous to the original paddies. SRF had produced more rice from those combined lands than had been the case before the revolution. The state farm, therefore, had contributed to the increase in national rice production after 1979. On the whole, SRF's rice-producing UPEs represented about seventeen hundred acres of irrigated land and were all grouped under Complex A, whose exclusive activity was rice production.

Complex A was managed by one administrator, who reported to SRF's director and who had a staff of six people: a foreman, an accountant, a supervisor, a payroll officer, a cook, and a tractor driver. Labor in the fields was divided among approximately fifty workers, male and female, and an irrigation team composed of about fourteen people. On average, therefore, Complex A employed about seventy people at all times. Wage data extracted from the original ninety-six biweekly individual payrolls of each of these seventy people

---

58. For an analysis of the structure of Nicaraguan state-owned agribusiness enterprises, see Austin and Krüger, "Food Policy."

Figure 6. The State Rice Farm (SRF) organization chart

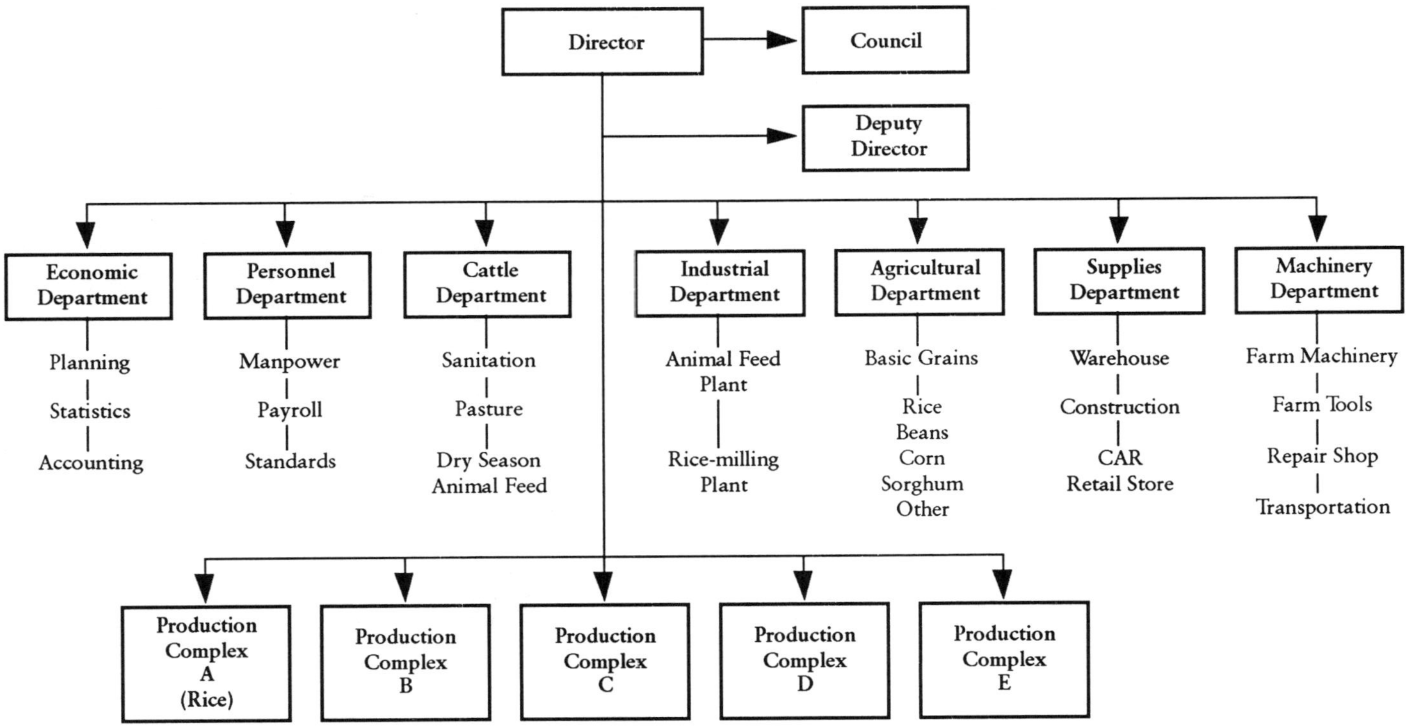

```
                              ┌──────────┐
                              │ Director │
                              └────┬─────┘
                                   │──────────────► Council
                                   │──────────────► Deputy Director
   ┌──────────┬──────────┬─────────┼─────────┬──────────┬──────────┬──────────┐
   ▼          ▼          ▼         ▼         ▼          ▼          ▼          ▼
Economic   Personnel   Cattle   Industrial  Agricultural  Supplies   Machinery
Department Department Department Department  Department  Department Department

Planning   Manpower   Sanitation  Animal Feed  Basic Grains  Warehouse   Farm Machinery
Statistics Payroll    Pasture     Plant        Rice          Construction Farm Tools
Accounting Standards  Dry Season  Rice-milling Beans         CAR          Repair Shop
                      Animal Feed Plant        Corn          Retail Store Transportation
                                               Sorghum
                                               Other
```

Production Complex A (Rice) | Production Complex B | Production Complex C | Production Complex D | Production Complex E

159

during the four-year period 1982–85 was pieced together with SRF's scattered records of daily wage rates paid in 1980 and 1981 to reconstruct the entire wage-rate history of Rice Complex A during the six-year period 1980–85. This history, presented in Table 23, shows that workers' total nominal monthly compensation increased by 182 percent in six years, from 1,237 cordobas in 1980 to 3,496 cordobas in 1985 (line 9). It also shows that two-thirds of this increase occurred in 1985 and that for all practical purposes, nominal wages during the first three years had remained unchanged.

By deflating this nominal wage by the consumer price index as calculated by the Central Bank of Nicaragua (BCN), we see that real monthly compensation had actually decreased from its starting level of 1,237 cordobas in 1980 to 247 cordobas in 1985 (Table 23, line 11). In other words, the purchasing power of cash wages paid to the workers of Rice Complex A had decreased by 80 percent in just five years. Somewhat paradoxically, the crop year 1984–85—the one in which the country's rice output began to fall—was for Rice Complex A both the year of the largest raise in nominal wages and the year of the largest decrease in real wages.

These measurements, however, should be taken with some caution. "Given the fact that Nicaragua does not have a specific consumer index of rural prices," explained a manager of the Central Bank, "and that the BCN consumer price index is primarily an urban index, it is probable that the actual fall in the purchasing power of rural wages is somewhat different. We can assume that food and housing would be cheaper in the countryside and that clothing and other urban products would be more expensive. By and large, nevertheless, our consumer price index still gives you the basic sense of what is happening."[59]

Within these limitations, an examination of the cost of the basic consumer basket over the period 1980–85 could shed some additional light on the predicament in which both workers and SRF appeared to have gradually found themselves. Table 24 shows the consumer basket cost for a typical Nicaraguan family of six (two adults, one adolescent, two children, and one infant) at official prices and at

---

59. Interview, 1985. This assessment appeared to be confirmed by the results of similar calculations using other available consumer price indexes. Deflating the wages of Rice Complex A by the consumer price index of the Nicaraguan Institute of Statistics and Census (INEC), for example, provided essentially the same results for the years 1980 to 1984 as did the BCN index. The only significant difference was in 1985, when, according to INEC, consumer price increases had been lower than those recorded by BCN and the corresponding loss in the purchasing power of SRF nominal wages would have been 68 percent instead of 80 percent.

Table 23. State Rice Farm: nominal vs. real wages of workers

		1980	1981	1982	1983	1984	1985
1	Daily wage	21.20	23.30	25.00	34.00	39.33	78.5
2	Basic monthly salary[a]	644.83	708.71	760.42	1,034.18	1,196.29	2,387.71
3	Plus social compensation						
4	Sundays	84.80	93.20	100.00	136.00	157.32	314.00
5	Vacations	53.00	58.25	62.50	85.00	98.32	196.25
6	"Thirteenth month"	53.00	58.25	62.50	85.00	98.32	196.25
7	Total monthly salary[b]	835.63	918.41	985.42	1,340.18	1,550.25	3,094.21
8	Plus per diem	402.00	402.00	402.00	402.00	402.00	402.00
9	Monthly nominal compensation	1,237.63	1,320.41	1,387.42	1,742.18	1,952.25	3,496.21
10	Consumer price index	100.00	133.40	163.00	216.60	325.30	1,412.90
11	Monthly real compensation (cordobas of 1980)	1,237.63	989.81	851.18	804.33	600.14	247.45
12	Index of real compensation	100	80	69	65	48	20

Sources: Payroll of State Rice Farm for wages and other compensation; and Central Bank of Nicaragua for consumer price index.
[a]Daily wage multiplied by 30.4167 days.
[b]The total of lines 2, 4, 5, and 6.

Table 24. Cost of the basic consumer basket at official and market prices (cordobas)

	Dec. 1980	Dec. 1981	Dec. 1982	Dec. 1983	Dec. 1984	Dec. 1985
At official prices						
Food	1,646	1,975	2,311	3,051	4,576	22,627
Clothing	805	966	1,130	1,390	2,281	8,540
Education	729	875	1,024	1,209	1,500	3,500
Health	578	694	819	917	1,200	3,000
Energy and water	200	240	281	332	412	1,149
Other	369	443	519	613	736	2,957
Total	4,327	5,193	6,084	7,512	10,705	41,773
At market prices						
Food	1,816	2,416	3,069	4,080	6,120	47,028
Clothing	1,338	1,794	2,207	2,726	4,490	15,698
Education	727	867	1,025	1,209	1,500	3,500
Health	519	700	820	967	1,200	3,000
Energy and water	150	150	150	150	150	1,500
Other	720	871	1,019	1,203	1,492	9,133
Total	5,270	6,798	8,290	10,335	14,952	79,859

Source: Central Bank of Nicaragua.

estimated black market prices. The Central Bank kept a record of both price series, since consumer basket goods were frequently unavailable at official prices, obliging consumers either to forgo the purchase or buy at higher black market prices. A comparison of Table 23 and Table 24 indicates that at no time between 1980 and 1985 was an SRF worker able to afford the basic consumer basket at official prices. The same was true even when both husband and wife worked and joined their salaries. Basic consumer basket goods at black market prices were obviously further out of reach.

If we look only at the cost of food, the situation is somewhat different. Table 25 reports the cost of the basic food basket as calculated by three different government institutions. By comparing Table 25 with Table 23, we see again that at no time could an SRF worker alone afford the basic family food basket. But if husband and wife both worked, joined their salaries, and spent all their combined income on food, they could afford the basic food basket at both official and black market prices during 1980 and 1981. In 1982, 1983, and 1984, they could still afford food at official prices, but black market prices became prohibitive. In 1985, their combined income could not afford the basic food basket at any price. In fact, their combined 1985 income of 6,992 cordobas represented, at best, only 43 percent of the cost of the basic food basket at official prices—using the most conservative cost index, as calculated by the Ministry of Internal Trade, as the point of reference. If the CIERA index of MIDINRA was chosen, the husband and wife's combined 1985 in-

Table 25. Cost of the basic food basket

Data source	1980	1981	1982	1983	1984	1985
1 Central Bank						
2 At official prices	1,646	1,975	2,311	3,051	4,576	22,627
3 At market prices	1,816	2,416	3,069	4,080	6,120	47,028
4 Ministry of Internal Trade[a]						
5 At official prices	1,287	1,382	1,616	2,144	2,629	16,244
6 At official and market prices[b]	1,402	1,548	1,753	2,285	3,298	17,310
7 CIERA[c]	1,255	1,471	1,711	2,120	3,021	19,560

Note: Costs are based on a family of six people: two adults, one adolescent, two children, and one infant.

[a]In May 1985 the Ministry of Internal Trade (MICOIN) changed the composition of the consumer basket.

[b]Market prices are used when products at official prices are unavailable.

[c]Part of the Ministry of Agriculture and Land Reform (MIDINRA).

come would have bought only 36 percent of the basic food basket. If the Central Bank index was used, the amount that could have been purchased would be reduced to 31 percent. One basic question remains to be answered: given the extremely harsh reality that these numbers reveal, how could workers have possibly survived?

The Survival Strategy: Switching from Farming to Trading

Some indication of the changes that occurred in Rice Complex A as the purchasing power of wages kept deteriorating can be inferred from the payroll information summarized in Table 23. First, the labor force switched from being predominantly male in 1982 to being mostly female in 1985. In Rice Complex A, the percentage of workdays performed by women rose from 11 percent in 1982 to 62 percent in 1985. On average, there were only three women workers in 1982 compared with thirty-seven in 1985. This phenomenon extended beyond Rice Complex A and was part of a broader trend in the countryside that was partly due to the recruitment of men into military service to fight the counterrevolutionary war. Interviews later revealed, however, that in the case of Rice Complex A, the rise in the female work force was primarily due to the particular survival strategies that workers developed, given that the farm was distant from the main cities and men did most of the traveling.

Second, between 1982 and 1984, women were never paid for overtime work. In 1985, payroll records show that women started working overtime, for a total of 640 workdays. As overtime work was paid at twice the normal salary, 640 overtime workdays represented more than one additional month of salary earned over the course of 1985 by each of the thirty-seven women, on average. Third, the average number of monthly workdays paid to female workers increased 38 percent, from 21 days in 1982 to 29 in 1985. For male workers, paid workdays increased 48 percent, from 21 days in 1982 to 31 in 1985. For the members of the irrigation team, whose payroll records were not available by gender, paid workdays increased 28 percent, from 35 days in 1982 to 45 in 1985. People could obviously work 31 days in a month at most; the discrepancy in the number of days worked was explained by the existence of an informal bonus system. Workers loyal to the company who did not quit their jobs were rewarded with "inflation adjustments" (*reajustes debido a inflación*). Members of the irrigation staff received higher salaries and were given more adjustments because their jobs were critical to the whole rice-farming operation. Fourth, the number of workers employed by Rice Complex A more than doubled between 1982 and 1985, from twenty-seven to sixty-one. For irrigation personnel, the number of people employed increased by 44 percent, from nine to thirteen.

An SRF supervisor explained:

> Our men are abandoning the farm. It is a real flight. Since they cannot make a living with what they earn, they do some *bisneo* [black market trading]. Every two weeks, when salaries are paid, the company sells each employee up to ten pounds of rice. The price is 12 cordobas per pound, which is quite good, since ENABAS [the state retail distributor] is selling it at 18 when available. The men take their rice and go to Managua, where they can sell it on the street at 190 cordobas per pound. In one day they can make a profit of 1,700 cordobas, which is what we pay them here for two weeks of work. If they do this once, all they think of afterwards is how they can do it again. When they come back here to work for the rest of the week, they don't put in the whole workday. They think about the next bisneo. We helped them earn overtime pay. But that was not enough. Even if you double the salary, it is not enough. Since rice is sold by SRF to each employee regardless of family ties, men put their wives and daughters to work here. But a young woman cannot put in as much work as an experienced laborer. Most of our work is highly mechanized. It takes time to learn how to use that equipment. Anyway, in some rice fields we have three sisters working together. Each one gets rice on payday, and their father sells it in Managua. This is not good. But if we don't help our workers' families, everybody will leave and we will be left without labor. Who would produce rice then?[60]

A MIDINRA manager from one of Managua's offices explained that the practice of black market rice trading was much broader than that.

> At the beginning, perhaps in 1982 or 1983, the black market for rice was somewhat sporadic and not systematically organized. Since 1985, with increasing inflation and price distortions, we have price controls on first-grade rice but no price regulations on broken rice. The result is that the market price for broken rice became three or four times higher than the controlled [official] price of first-grade rice. Private producers then began breaking their whole rice so that they could sell it as second-grade. That created a shortage of rice at official prices—first-grade rice was suddenly hard to find—which fed the black market and created incentives on the

60. Interview, June 1985.

part of rural wage earners to join the trade by keeping one foot in a farm and another in the city.[61]

At the same time, as an SRF manager clarified:

The company tries to do everything in its power to retain its workers and to keep production in the rice fields. We give our workers loans, we try to help them build their houses. When we have good construction wood, like the wood from the crates of the water pumps that we import, we give it away to the workers who perform better, so that they can make improvements in their homes. We help them get supplies from the rural state store [*centro de abastecimiento rural*, or CAR]. The company has its own store in the farm. We try to give them salary incentives for punctuality, discipline, good behavior. But the workers don't want money. They would prefer payment in kind. They say that even if we double their salaries, it won't buy the food they need. We can't compete with the black market.[62]

Explained a worker in Complex A:

I have worked in this company for several years. My wife and I have ten children. Before 1983, my wife and my elder daughter used to join the salaries that they earned at the farm with mine and that was enough to buy food for the family. Since then, things have become complicated. The company store now has no corn, no sugar, no soap, no cooking oil, no clothing. I can't afford to buy a new pair of pants any more. We cannot wash our clothing. Soap is too expensive on the free market. Our children go around barefoot. I don't keep track of the price of meat any more, since it's beyond our reach. The company has tried hard to help me build the walls around our roof. But I need cement and bricks, and they don't have it. It's too expensive to buy it elsewhere. What my wife and I do now is sell onions and other fresh vegetables in the city, while two daughters and my young boy work at the farm. We buy the pound of onions at 60 cordobas, and we can resell it in Managua at 150, sometimes at 200. On a

---

61. Interview, June 1987. The presence of a black market for rice also created many problems for private producers not involved in it. "The Ministry of Internal Trade [MICOIN]," explained Mario Hannon, president of the Association of Rice Producers [ANAR], "is almost terrorizing us. I am a rice producer, and my rice mill was making a truck delivery of six hundredweights of rice, one for me, four for some friends of mine, and one for an administrator that is also a friend of mine. MICOIN confiscated all that rice and I had to thank the ministry for not having confiscated my truck too. When I went to the ministry to complain about my six hundredweights of rice, they replied to me by saying: 'You should be grateful to us that we didn't confiscate your vehicle too. The law says that we should have confiscated your vehicle too'" (INIES, *Foro-debate*, no. 1 [Managua: INIES, 1987], p. 37).

62. Interview, 1985.

good day we can earn more than 1,000 cordobas. I wish we could do that every day. But sometimes we can't find onions to resell, and sometimes we don't find transportation to the city. Besides, my wife has to take care of the children at home, too. And we still work at the farm, but part time. We can buy rice there every other week.[63]

The Fall in Labor Productivity in Rice Complex A

What implications did this situation have for labor productivity? Informal interviews with SRF supervisory and labor personnel confirmed that the length of the workday had decreased in 1984 and 1985 as workers tried to perform more than one job at a time and new, less experienced personnel were hired to replace increasingly absent male workers. SRF company records did not, however, report the number of hours of work performed by each laborer in one day, so it was not possible to quantify the decrease. The payroll of Rice Complex A indicated that workers were always paid their full daily wage, regardless of the actual number of hours they worked.

Other indications of a fall in labor productivity were available, however, and are reported in Table 26. First, it can be seen by matching the payroll against the acreage planted that the ratio of workdays per manzana increased from 13 days in 1982 to 23 days in 1985, a 77 percent increase in labor input per area planted. This increase was particularly alarming considering the fact that the labor standard set by MIDINRA for the cultivation of one manzana of irrigated rice was 8.12 days of labor. Between 1982 and 1985, SRF had never met this standard, and as workers' real wages decreased, so did the amount of work they performed in one day.

Second, by comparing the payroll to rice output (Table 26), it can be seen that between 1982 and 1985, the amount of rice produced in one day of labor was cut in half. In 1982, one day of labor produced 4.07 hundredweights of rice. In 1985, it produced only 2.10 hundredweights.

The largest losses in labor productivity, both in terms of output per workday and workdays per unit of land, occurred in 1984 and 1985, when the drop in the purchasing power of wages was greatest. Total rice output for Complex A in the agricultural cycle 1984–85 was also lower than in 1983–84.

63. Ibid.

Table 26. State Rice Farm (SRF) Complex A: labor input and rice output

		1982	1983	1984	1985
1	Average no. of workers	27	35	73	61
2	Average size of irrigation team	9	20	12	13
3	Total (lines 1 + 2)	36	55	85	74
4	No. of days of labor	10,725	16,573	24,501	29,434
5	Area planted (manzanas)	830	1,199	1,141	1,285
6	Rice output (cwt milled)	43,714	66,292	61,213	61,975
7	Days of labor per manzana planted	13	14	21	23
8	Rice output per one day of labor	4.07	4.00	2.49	2.10

Source: State Rice Farm payroll and production statistics.

"If we hadn't hired more people, even inexperienced people, our rice production would have fallen even more," explained an SRF manager. "Of course, we cannot increase the salaries as the workers would like. Many of them now think in terms of black market prices. But the company doesn't decide what the wage rate should be. We must follow the government's regulations. The fact remains that we have done everything in our power to retain our workers. And our workers know it. Now, if our best mechanic leaves us because a black market repair shop in Managua has offered him more than ten times our salary, whose fault is it if, as a result, the water pump is not working properly, the tractor breaks down, and the water dam leaks?"[64]

Wage Policy, Inflation, and the Growth of the Black Market

Government studies analyzing the entire economy indicated that much of what was occurring in SRF was also happening elsewhere in the state sector on a generalized scale. If the private sector had been able to avoid in part this drop in labor productivity, it appeared to be because it had been more active in the black market, where large monetary gains allowed its participants to operate on a different cost structure, including the cost of wages.[65]

Typical was the case, reported by MIDINRA, of some private rice

64. Ibid.
65. Interview with government officials, 1987.

producers or distributors that withdrew their first-grade rice from the official market at regulated prices, then paid a rice mill to convert it into second-grade broken rice, and finally sold it legally through nonregulated channels at four times the price of higher-quality rice.[66] These rice suppliers were operating in a different economy. The monetary incentives available in that environment were powerful enough to make wage earners in the state farm sector leave their jobs and change their line of activity, from producers in the formal economy to distributors in the unregulated market. In some cases the incentives were so strong that whole families migrated to be closer to the new activity.[67] Managua was the typical destination.[68] An extreme, yet illuminating example was the related case of the state-owned enterprise AGROMAQ, which had responsibility for the sale and maintenance of rice irrigation machinery and other farm equipment in the agricultural sector. A company study revealed that personnel turnover among its approximately six hundred workers was over 100 percent per year. The firm lost its entire labor force every ten months. One-fourth of those who left did not notify the company of their resignation and appeared on company records as having abandoned their workplace.[69] The network of private garages and farm equipment repair shops operating at black market prices could obviously offer higher remunerations to a worker with technical skills than could the state company.[70]

On a smaller scale, the same phenomenon was happening on SRF, where a highly mechanized and irrigated system of cultivation made its best technical people vulnerable to similar pressures. It was illegal to raise wages above government-regulated rates; a typical way to get around the regulation was to pay part of the salary in kind. In the case of a repair shop, this payment in kind might have been a few tractor spare parts, which could then be resold on the black market at prices often exceeding several times the nominal cash wage. SRF countered those offers by selling rice to their employees at favorable prices. In other state-owned rice farms, the amount of rice given biweekly by the state company to each of its employees was as high as twenty-five pounds, and it was given free of charge.[71] Sub-

66. Interview, June 1987.
67. UNRISD, "Urbanization and Food Systems Development," pp. 190–201.
68. Ibid.
69. Empresa de Maquinaria Agrícola, R.A., *Evaluación de los indicadores económicos de la empresa* (Managua: AGROMAQ/MIDINRA, 1987), p. 16.
70. Interview with an AGROMAQ manager, 1986.
71. Interviews, 1985 and 1986. See also *Nuevo diario,* February 2, 1987.

sequent investigations revealed cases in which rice was systematically stolen from state enterprises to be sold on the black market. "Rice reaches the black market not only by way of payments in kind, as is practiced in state rice farms," explained the progovernment newspaper *Nuevo Diario,* "but also through leaks in the warehouses of [the state-owned grain distributor] ENABAS, and through disappearances in the official state retail network."[72]

Data from the Central Bank reveal how accurately the wage phenomenon found at SRF reflected the national agricultural experience. There is a striking similarity between the dichotomy of nominal and real wages within SRF and that found on the national level. An independent study calculated that, on average, real wages in Nicaragua had lost approximately 60 percent of their purchasing power between 1980 and May 1985. This figure would have been even closer to that presented for SRF if allowance had been made for the additional inflation occurring in the second part of 1985.[73]

In the same vein, a comparison between the cost of the basic consumer basket and the national system of mandatory wage scales introduced by the government in 1984, commonly called SNOTS, showed, for example, that in December 1984 only 4 percent of all wage earners could afford to pay the basic consumer basket. Only a small fraction of them, that is, an estimated five hundred people—less than one-fifth of 1 percent of Nicaragua's wage earners—were able to afford the basic consumer basket at market prices.[74] These five hundred people belonged to the SNOTS scale 28 and represented those who earned the highest permissible salary in Nicaragua. In the state sector, these people consisted of the highest members of the public administration. In the event of supply shortages in the official retail channels, anyone below these levels could not afford the full basic consumer basket at market prices. The result was that many middle- and low-level state administrators frequently spent part of the day in line at some state store, making sure not to miss their quota of the basic consumer basket at controlled prices.[75]

This overall situation was summarized by an economist of MIDINRA in the following terms:

---

72. "Payments in Kind and Thefts in Warehouses: Rice Reaches the Black Market in the Form of Many Small Amounts," *Nuevo Diario,* February 2, 1987.
73. Stahler-Scholk, *Política salarial,* p. 13.
74. BCN.
75. Interviews, 1984–1985.

> Inflation is not caused by speculators [as the Ministry of Internal Trade and the government press often claimed]. These black market operators simply profit from the inevitable price differential that results from a policy of price controls. An excessive increase in the money supply, together with other factors that influenced the demand for foreign currency, increased the black market exchange rate which, together with a shortage of goods, raised the price level in the parallel [black] market. Price controls in the official market proved to be unenforceable in an environment where demand was stimulated by too much money printing and the supply of goods was shrinking. . . . This [price control] policy has negatively affected wage earners because they are obliged to sell their labor force in a price-controlled market—the market of wages regulated by the SNOTS system—while at the same time they have to purchase part of their consumer goods in the parallel [black] market, where prices are determined by the law of demand and supply. Since controlled wages do not allow workers to live, workers have joined the so-called informal [black market] sector of the economy.[76]

This generalized shift of the labor force away from the formal sector, along with the nature of the forces that prompted it, were confirmed by a number of government reports, including the 1985 Economic Plan, which stated:

> [Among] the phenomena that demonstrate with more clarity the deterioration of the economic situation are the following: a marked decrease in the supply of goods and services to the population, which becomes more acute starting in 1982; . . . an inflation that has been growing year after year, which affects in a fundamental way the income level of wage workers and discourages labor from joining activities in production and essential services, such as health; . . . a situation of chaos in the pricing system, which introduces irrationality in the allocation of products, discourages production, and promotes speculation and smuggling; . . . a drastic drop in average labor productivity, particularly in the agricultural sector, which causes cost increases and a decrease in production.[77]

A United Nations study of the Nicaraguan food system completed in 1986 further clarified the implications of the government's pessimistic overall assessment, stating:

> Rapidly increasing inflation and a restrained wages policy has caused the real wages, particularly of agricultural workers, to fall in recent years. As

---

76. José Luis Medal, *La crisis y las políticas macroeconómicas* (Managua: CINASE, 1986), p. 21.

77. SPP, *Plan económico 1985*, p. 3.

a consequence, the ranks of the commercial sector have been swelled not only by migrants from the rural areas, but also from industry. Migration and the expansion of the urban informal sector has contributed to what has become one of the major constraints on the development process, namely labor shortages, particularly in agriculture and agroindustry. This affects not only production but also the implementation of an ambitious investment program which the government has promoted to boost export and basic food consumption.[78]

Conclusion

We have seen in this chapter that the persistent deterioration of labor productivity throughout the first six years of the postrevolutionary government jeopardized the viability of the state sector as the main generator of economic surplus, which had been central to the vision of a "new Nicaraguan economy." Despite repeated government efforts to redress the situation, by 1985 it had become clear that labor productivity had not improved after the expected drop in productivity due to a postrevolutionary labor adjustment phase. At the root of the problem was a powerful internal dynamic triggered by the simultaneous presence of two phenomena. One was the dramatic loss in the purchasing power of wages: particularly after 1982, work on state farms was viable economically only insofar as it provided a base for a set of survival strategies, in which the state company as employer had an increasingly marginal role. The other was a highly distorted food price structure, which offered a ready-made solution to wage earners seeking additional income.

Aware of these phenomena, managers of state-owned enterprises struggled to retain their workers and to motivate them by offering the best material incentives possible short of violating mandatory wage guidelines. They were unable, however, to close the income gap created by a social wage that, as a matter of policy, divorced cash wages from the inflation rate. The overall outcome was a transfer of human resources away from state production into commercial activities in the informal economy, at the further cost of labor productivity.

Up to this point, state-owned agribusiness enterprises have been shown to have sustained substantial financial losses, to have operat-

78. UNRISD, "Urbanization and Food Systems Development," p. 202.

ed with considerable idle plant capacity, and to have struggled with decreasing labor productivity. These problems were largely stimulated by adverse macroeconomic policies in the area of foreign exchange rates, domestic food prices, and wage rates. Given the magnitude of the difficulties encountered, one might well ask, How could state enterprises have possibly endured such pressures for so long? How were state enterprise managers able to keep production going, despite the odds against them? Why didn't they run out of cash? Why didn't they go bankrupt? How were they able to avoid closing down? The next chapter—an analysis of credit policy and the problem of the indebtedness of state-owned enterprises—explores these questions.

6

Credit

> With inflation at 400 percent and interest rates at 20 percent, it is more expensive for us to buy a carburetor for our tractor in the black market than it is to buy a brand new tractor with borrowed money from the state dealer. So, we buy the new tractor and let the other one rust in the repair shop.
> —Manager, state agribusiness enterprise, 1985

> No one in the government would let a state enterprise run out of money. We are forced to approve any loan that these firms request. However, nobody cares if in the process the [state] banks go broke.
> —Government official, National Financial System, 1985

One of the first decisions made by the revolutionary government of Nicaragua when it assumed power in July 1979 was to nationalize all private domestic financial institutions. The decision, formalized in August of that year, aimed to strengthen the banks, whose financial situation had been damaged by large transfers of assets during the months preceding the fall of the Somoza regime.

Once the banks were nationalized, the government proceeded to streamline the entire structure of the country's financial system.[1] In mid-1980, through a series of mergers, the number of financial institutions was cut in half. Out of that reorganization were left the Cen-

---

1. Little government documentation exists on the structure of the Nicaraguan banking system after 1979. Most information in this section was obtained from personal interviews with bank executives in 1986 and 1987. A description of the Nicaraguan banking system can be found in country reports made by the World Bank and the International Monetary Fund (See in particular, IMF, *Nicaragua: Recent Economic Developments* [Washington, D.C.: IMF, 1986]; IMF, *Country Program Paper* [Washington, D.C.: IMF, 1982]; World Bank, *Second Agricultural Project* [Washington, D.C.: World Bank, 1981]; and World Bank, *Nicaragua: The Challenge of Reconstruction* [Washington, D.C.: World Bank, 1981]).

tral Bank, five domestic banks, four nonbank intermediaries, and three branches of foreign banks.

The five domestic banks were the Banco de America (BAMER), the Nicaraguan Bank (BANIC), the National Development Bank (BND), the Popular Credit Bank (BCP), and the Banco Inmobiliario (BINMO). Credit specialization by economic sectors was encouraged, and BND, assigned to serve the farm sector, rapidly emerged as the largest Nicaraguan bank, accounting in 1980 for 70 percent of the agricultural portfolio and 53 percent of the entire portfolio of the banking system. The four nonbank intermediaries included the National Investment Fund (FNI), which was created to administer long-term investments and centralize all foreign borrowing for development purposes, and three largely inactive mortgage agencies. The three branches of foreign banks belonged to Citibank, the Bank of America, and the Bank of London. After 1979, these branches were not allowed to accept deposits from local sources. This and other restrictions reduced their activities in the domestic market to a purely nominal presence.

The resulting new banking structure, designated the National Financial System (SFN), was limited to a few specialized, well-established institutions that were fully backed by the government. It had been designed with the stated intent of cutting administrative costs, using scarce trained personnel more efficiently, and increasing public confidence in the banking system. It was placed under the control of the Nicaraguan Financial Corporation (CORFIN), a newly created holding company providing overall guidance in the implementation of government monetary and credit policy. CORFIN, in turn, reported directly to the Central Bank of Nicaragua (BCN).

As the task of reactivating the economy became a primary concern of the new government, credit, together with support prices, was chosen as the leading policy instrument to stimulate production. State-owned agribusiness enterprises gradually became the major beneficiary of this policy, and as a result, the country's economic prospects became increasingly dependent on the relationship between the SFN and the APP. Between 1980 and 1985, that relationship developed into a great source of tension within the government. The state enterprises' share of the country's total credit increased from 28 percent in 1980 to 63 percent in 1985 (Table 27), and as their debt increased, so did their financial losses. Unable to service their debt, state enterprises called for a restructuring of bank loans and more borrowing. In 1982 the SFN estimated that approximately

Table 27. Share of credit by economic sector (%)

Credit	1980	1981	1982	1983	1984	1985
Agricultural credit		100	100	100	100	100
APP[a]	N.A.	36	39	44	51	61
Private sector	N.A.	64	61	56	49	39
Livestock credit		100	100	100	100	100
APP[a]	N.A.	26	35	41	44	47
Private sector	N.A.	74	65	59	56	53
Industrial credit[b]		100	100	100	100	100
APP[a]	N.A.	47	57	73	79	83
Private sector	N.A.	53	43	27	21	17
Total credit[c]	100	100	100	100	100	100
APP[a]	28	33	40	45	55	63
Private sector	72	67	60	55	45	37

Source: National Financial System (SFN).
[a]Includes all state-owned enterprises.
[b]Includes agroindustry.
[c]Includes commercial and all other credit.

half the agricultural loan portfolio of the state enterprises was in default (Table 28, line 6). In the same year, the total agribusiness loan portfolio held by the state enterprises, including agroindustries and trading operations, was about 40 percent in default. As a result, the banks were unwilling or unable to finance further defaults.

Despite these problems, the government deemed it critical to continue to support state enterprises. Three factors prompted this attitude. First was an overwhelming desire to increase the country's physical output. Because of the disruptions of the revolution, gross domestic product in 1979 had been, in real terms, nearly 27 percent less than in 1978, and the new government could not achieve its food policy goals without raising domestic output to prerevolutionary levels. Later, it even needed to surpass them. Given the large potential production capacity of state enterprises, the government could not afford to let them go under for lack of financing. If the country was to recover from the disruptions of the revolution, it could do so only if the state enterprises were part of that recovery.

Second, in 1982 the government was still convinced that the financial problems of state enterprises were transitional and organizational in nature. It was thought that improved control systems and the development of more experienced managers would suffice to redress these problems. Therefore, the government reasoned, the enterprises deserved a second chance, and that chance would be provided by additional financing.

Table 28. Credit given to state-owned enterprises (APP) and estimated amounts in default, 1980–1985 (millions of cordobas)

Credit		1980	1981	1982	1983[a]	1984	1985
1	Total credit to APP[b]	3,395	4,671	6,673	10,394	14,110	27,022
2	Amount in default[c]	N.A.	1,247	2,491	2,303	3,082	5,183
3	Percentage in default	N.A.	27	37	22	22	19
4	Credit to state agriculture	N.A.	1,653	2,169	2,692	4,121	9,341
5	Amount in default[c]	N.A.	520	1,017	717	869	1,756
6	Percentage in default	N.A.	31	47	27	21	19
7	Credit to state livestock	N.A.	447	997	1,661	2,186	4,516
8	Amount in default[c]	N.A.	54	109	158	198	225
9	Percentage in default	N.A.	12	11	9	9	5
10	Credit to state industries[d]	N.A.	2,330	3,192	5,703	7,455	12,007
11	Amount in default[c]	N.A.	633	1,310	1,377	1,863	2,887
12	Percentage in default	N.A.	27	41	24	25	24
13	Credit to state trading[e]	N.A.	139	189	280	290	1,054
14	Amount in default[c]	N.A.	40	55	51	95	212
15	Percentage in default	N.A.	29	29	18	33	20
16	Other credit to APP[b]	N.A.	241	126	58	58	104
17	Amount in default[c]	N.A.	—	—	—	57	103
18	Percentage in default	N.A.	—	—	—	98	99

Source: National Financial System (SFN).
[a]1983 is the year of the first financial bailout (*saneamiento*) of the state-owned enterprises.
[b]APP, or "Area of People's Property," includes all state-owned enterprises.
[c]Loans over 90 days in arrears. Amounts exclude interest payments.
[d]Includes agroindustry.
[e]Includes such state trading enterprises as ENABAS.

Third, between 1979 and 1985 the government had not yet given up on its plan to make state enterprises the center of the country's new model of capital accumulation. According to MIDINRA's major document on the country's agricultural development strategy, *Marco estratégico del desarrollo agropecuario*, in the year 2000 state enterprises would have represented 30 percent of the country's farm output, that is, almost double what they had represented during the period 1980–82 (see Table 8). Given all this, it was unlikely that the government would easily abandon state enterprises.

Motivated by these factors, in 1983 the government implemented a financial bailout (*saneamiento financiero*) in an effort to save the state enterprises. Soon after, it increased its efforts to preserve the newly endowed capital structure. By 1985, though, state enterprises were more indebted than they had been in 1982, and the accumulation of loans in arrears once again threatened the stability of the financial system. The government thus felt compelled to negotiate yet another bailout. This turn of events was particularly disturbing to policymakers, because it called into question the feasibility of the so-called new model of state accumulation, a central component of the New Economy. Why had the state enterprises been such a persistent drag on the country's financial resources?

This chapter examines the expansionary credit policy that was sustained by subsidized interest rates and the facilitating credit decision-making mechanisms specially designed for the state enterprises. It suggests that these proved to be an irresistible vehicle for massive indebtedness for many state firms afflicted by financial problems, low labor productivity, and idle plant capacity. This massive indebtedness appeared to be justified to state enterprise managers in light of evidence that financial difficulties and inefficiencies had become largely exogenous variables about which little could be done, short of altering or contravening the government's macro price policies. In addition, the administrative culture that ensued from these managerial constraints called for production without regard for balancing the relationship between costs and revenues, which removed whatever inhibitions to further indebtedness still remained among state enterprise managers. Thus, the generous credit policy acted as a temporary palliative to an impending financial crisis. Perhaps more important, it perpetuated inefficient production and further postponed the need to revise foreign exchange and domestic price policies.

The Financial Structure of the State Sector: Goal and Performance

In 1979 the revolutionary government of Nicaragua lacked a clear vision of the financial structure needed by its new state enterprises for the immediate future.[2] At that time the administration was already overextended just trying to identify exactly what Somoza and his associates owned, what it should expropriate, and where the

2. This section is based on interviews with government officials in various ministries, as well as on the following documents: MIDINRA, *Estados financieros: Junio 1981* (Managua: MIDINRA, 1981); MIDINRA, *Balances generales consolidados de las regiones al 30 de junio de 1981* (Managua: MIDINRA, 1981); *Acuerdo: Junta de Gobierno de Reconstrucción Nacional de la República de Nicaragua* (Managua: November 2, 1982); *Acuerdo: El Ministerio de Desarrollo Agropecuario y Reforma Agraria* (Managua: March 1983); Walter Krüger and James E. Austin, *Organization and Control of Agricultural State-owned Enterprises: The Case of Nicaragua* (Boston: Harvard Business School, 1983); and MIDINRA, *Balances generales consolidados al 31 de marzo de 1985* (Managua: MIDINRA, 1985).

over one thousand confiscated farms were that represented the initial endowment of the state sector.[3]

The new policymakers were more definitive on two closely related issues. First, the foundation had been set for the state to become the center of the country's capital accumulation. This goal was accomplished through the expropriation of Somoza's major sources of profit, as well as those of his associates, and through a corresponding transfer of these assets to the state enterprises. Second, the future growth of the state enterprises would have to be financed by the enterprises themselves, given a reasonable adjustment period.

Opinions differed among the various government branches as to the length of the adjustment period before the achievement of economic surpluses and self-financing. The consensus was that the adjustment period ought to be as short as possible, for the sake of consolidating the revolution and getting on with the new model of state profit accumulation. Nevertheless, those government branches directly responsible for state enterprises, such as MIDINRA, had somewhat lengthier estimates than those who did not, such as the Ministry of Planning. In the climate of optimism prevailing within the government in 1979, an estimate of two years was adopted. Reflecting this assessment, the government's first economic plan set an overall output goal for 1980 equal to the prerevolutionary level of 1978. For 1981, government output targets aimed at the "normal" and higher level of 1977.[4] After 1981, accumulation of state profit would begin, and state enterprises would function as the engine of growth for the entire economy.[5] Meanwhile, a special State Accumulation Fund would be created for the purpose of centralizing the profits of state enterprises and establishing a mechanism whereby future investments could be centrally planned.

The magnitude and unexpected nature of the financial losses generated by state enterprises after 1979 has already been discussed in Chapter 3. In a similar fashion, the heavy indebtedness of the state enterprises and their inability to meet financial obligations on time also caught the government by surprise. It was only in 1981 that the problem was fully recognized within government circles. That year, a MIDINRA survey indicated that the debt contracted by the state

---

3. For an analysis of the organizational challenges of the APP in its first years, see Krüger and Austin, *Control of Agricultural State-owned Enterprises*.
4. See, for example, Ministry of Planning, *Programa de reactivación económica en beneficio del pueblo 1980* (Managua: MIPLAN, 1980), p. 17.
5. Interviews with government officials, 1985 and 1986.

enterprises—mostly in the form of short-term obligations—represented approximately 74 percent of their total assets and tended to increase rapidly as a result of operating losses. The financial bailout that followed the survey forgave about one-third of the debt that state enterprises had with the National Financial System and converted the balance into long-term obligations. In this transaction, the debt of the state enterprises was reduced to 58 percent of total assets. For the first time, the government specified what would constitute a sound financial structure for state enterprises. The bailout agreement was based on the understanding that the new ratio of debt to total assets represented a financial leverage consistent with the actual loan repayment capability of state enterprises. Accordingly, the agreement stated that the financial structure resulting from the bailout henceforth rendered possible a fair evaluation of the economic performance of the state enterprises. This statement was meant as an implicit acknowledgment that the adjustment period had ended and that state enterprises would subsequently be held responsible for their financial soundness.

Concurrent with the implementation of the bailout, an extensive program was initiated to improve the planning, budgeting, and financial control capabilities of the state enterprises. Special attention was placed on developing the administrative capabilities of state enterprise managers, particularly in finance.[6] In spite of significant improvements in these areas, however, state enterprises continued to rely on massive indebtedness. By 1985, state enterprises' renewed difficulties in meeting their financial obligations prompted MIDINRA to conduct a new survey. This study indicated that (1) total indebtedness had increased threefold between 1982 and early 1985; (2) these firms, by March 31, 1985, had become more leveraged than they had been before the 1983 bailout; and (3) additional indebtedness was expected in the foreseeable future.[7]

Why did state enterprises continue to rely on massive borrowing during their six years of existence despite repeated government efforts, particularly after 1982, to maintain a more sound financial structure? To answer this question, we now turn to the analysis of

---

6. For an analysis and evaluation of the Nicaraguan experience concerning the development and training of managers in state enterprises, see James E. Austin and John C. Ickis, "Management, Managers, and Revolution," *World Development* 14, no. 7 (1986): 775–790.

7. MIDINRA, *Balances generales consolidados al 31 de marzo de 1985*.

credit policy in the context of a state mill operating in the sugar agribusiness commodity system.

The Indebtedness of State Enterprises: The Case of the State Sugar Mill SSM

The Dimension of the Problem

The sugar mill SSM is one of Nicaragua's six state-owned sugar mills and one of seven sugar mills in the country (Figure 7).[8] Before its expropriation in 1979, this mill was owned by Luis A. Somoza D. y Cia. Ltda., a company controlled by the Somoza family. In 1986, during a review of the company's control systems with SSM staff, the original financial records of the Somoza company for the years 1974 through 1977 were fortuitously discovered. These records provided a rare comparative view of the financial structure of SSM before and after the revolution. Most private agribusiness firms that had been expropriated in 1979 and converted into state-owned enterprises lacked such documentation. In most cases, prerevolutionary records had either been lost or destroyed. Conversion of postrevolutionary financial records for 1981–85 from MIDINRA's SUCA (System of Uniform Administrative Control) control system into conventional financial categories produced a financial track record of nine years.

From these data emerged two basic points regarding the profile of the company's financial structure and the evolution of its level of indebtedness. First, the debt ratio, which measures the percentage of total funds provided by creditors, went from 29 percent in 1974 to 116 percent in 1985. During the prerevolutionary years for which data are available, total liabilities had never exceeded 52 percent of

---

8. This section is based on (1) interviews with major participants in the Nicaraguan sugar agribusiness system; (2) three months of research with the assistance of SSM management; (3) access to SSM internal documentation; (4) unpublished government documents (among which are MIDINRA, *La agroindustria de la caña de azúcar en Nicaragua* [Managua, 1986]; MIDINRA, *Incentivos implementados en el sector azúcar* [Managua: MIDINRA, 1984]; MIDINRA, *Estudio de factibilidad Proyecto Agroindustrial Azucarero Tipitapa-Malacatoya* [Managua: MIDINRA, 1981]; MIDINRA, *Propuesta de organización de la empresa comercializadora de azúcar* [Managua: MIDINRA, 1987]; MIDINRA, *Los problemas de la comercialización interna del azúcar y sus alternativas de solución* [Managua: MIDINRA, 1987]); and (5) published nongovernment documents (among which the most informative were Alfonso Dubois et al., *El subsistema del azúcar en Nicaragua* [Managua: INIES, 1983]; José Luis Coraggio, *Posibilidades y dificultades de un proyecto regional alternativo: El caso del subsistema del azúcar de caña* [The Hague: ISS, 1983]; and Alfonso Dubois, *La economía mixta en transición: El caso del azúcar* [Managua: INIES, 1983]).

Figure 7. The structure of the Nicaraguan sugar commodity system: estimated state share of output, 1983

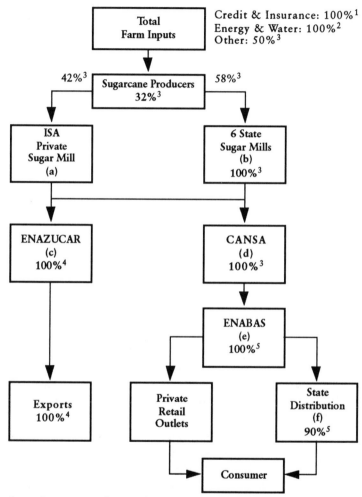

Sources: [1]BCN; [2]INE; [3]MIDINRA; [4]MICE; [5]MICOIN.
(a) ISA is the Ingenio San Antonio sugar mill.
(b) The six state mills are Julio Buitrago, Germán Pomares, Benjamín Celedón, Camilo Ortega, Victoria de Julio, and Javier Guerra.
(c) ENAZUCAR is a state company that is the sole exporter of Nicaraguan sugar.
(d) CANSA was nationalized in July 1981; until then, it was owned by ISA.
(e) State enterprise that procures and distributes sugar and other basic foodstuffs.
(f) In February 1982 sugar began to be rationed through state distribution channels. Priority was given to sales to the army, the Ministry of Interior, hospitals, and state distribution networks such as CAT, CAR, and ERCAI.

total assets and, on average, had represented 46 percent of total assets. After the revolution, total liabilities were never less than 85 percent of total assets. This percentage occurred in 1981; in 1982 the debt ratio went up to 98 percent. In 1983 it decreased slightly to 90 percent, but only because at the beginning of that year, SSM, a state-owned company since 1979, had benefited from the government's financial bailout. In that operation, the government forgave eighty-six million cordobas of SSM debt and injected sixty-six million cordobas of additional common stock into the company. Without the bailout, the funds provided by creditors would have exceeded total assets, and the company would have been bankrupt. The bailout, however, did not prevent SSM's total liabilities from rising to 101 percent of total assets in 1984. In 1985, total liabilities reached 116 percent of total assets.

Second, most of the increase in SSM's liabilities came from short-term debt that, according to Nicaragua's accounting classifications, was scheduled for repayment within eighteen months. On average, during the prerevolutionary years 1974–77, short-term debt was 15 percent of total assets and never exceeded 22 percent of total assets. After the revolution, however, short-term debt averaged 77 percent of total assets and rose from 64 percent in 1981 to 98 percent in 1985. In 1985, total current liabilities alone exceeded total assets by a margin of 8 percent. For all practical purposes, after 1982 the state sugar mill SSM was bankrupt. "We have many state-owned companies in that situation," explained an executive of the National Financial System in discussing SSM's financial statements. "So, the banks are going bankrupt too. In order to continue lending to state enterprises, we have to borrow from the Central Bank, which in turn has to keep issuing more money."[9]

Although the U.S. quota on Nicaraguan sugar had been reduced during the postrevolutionary period and was discontinued, along with all other U.S. import-export transactions, following the 1985 trade embargo, the Nicaraguan sugar export price rose from its average of $0.148 per pound between 1974 to 1977 to an average of $0.175 per pound between 1980 and 1983. This amount represented an average price increase of 18 percent. The Nicaraguan sugar export price, therefore, did not seem to have been directly responsible for SSM's 1981–83 massive indebtedness and bailout experience. What, then, was behind this phenomenon?

9. Interview, fall 1986.

The Pressure to Borrow

Four major areas that pushed the company into heavy borrowing were identified through an analysis of SSM's financial records, as well as through discussions with the management of the company and with various participants of the Nicaraguan sugar agribusiness commodity system, including the state bank that financed SSM. These areas related primarily to the need to finance rapidly growing accounts receivable, inventories, and fixed assets at a time when the company's net worth was shrinking because of financial losses.

First, between 1981 and 1985 there was an almost tenfold increase in accounts receivable generated from ordinary sugar sales operations, from 35 million cordobas to 337 million. Meanwhile, sales during the same period had increased only threefold (Table 29). The average collection period had increased from 153 days in 1981 to 474 days in 1985. The prerevolutionary average collection period was 78 days. What was the cause of this increase?

"The cause is very simple," explained an SSM manager. "ENABAS [the state company in charge of purchasing and distributing all sugar in the domestic market] sometimes waits more than one year before paying us for our deliveries."[10] While SSM was part of MIDINRA, ENABAS reported to the Ministry of Internal Trade (MICOIN) (Figure 7). Since ENABAS knew that SSM, as a state enterprise, had practically unrestricted access to credit—as we shall see in the next section—it chose to postpone paying for SSM sugar deliveries. By waiting one year, ENABAS not only improved its cash flow at no financial cost but also benefited from a substantial depreciation of its accounts payable, thanks to double- and triple-digit inflation. Since it was a state company just as was SSM, ENABAS was able to argue its position by saying that its own customers were postponing payment to it. Indeed, the balance sheet of ENABAS indicated that its own accounts receivable were about 77 percent of total assets during the period 1981-85: the customers of ENABAS were not paying their bills either. Sugar mills, however, were required by law to deliver their sugar to ENABAS, regardless of payments. Because of a policy that made credit readily available to state enterprises, it was much easier for SSM, as well as for the other state-owned sugar mills, to get an additional loan to finance overdue accounts receivable than it was to get ENABAS to improve its payment schedule.

---

10. Interview, fall 1986.

Table 29. State Sugar Mill (SSM): income statements (millions of cordobas)

Claims on assets	1974	1975	1976	1977	1981	1982	1983	1984	1985
Net sales	33.00	61.24	89.58	62.55	84.01	103.68	144.85	338.64	259.86
Costs of goods sold	15.56	33.92	55.13	38.22	51.25	93.31	129.21	306.66	250.02
Gross profit	17.43	27.32	34.45	24.33	32.76	10.37	15.63	31.99	9.84
Less operating expenses:									
Selling	6.48	7.28	6.44	7.04	13.57	3.87	1.31	6.32	6.51
General and administrative	2.20	3.57	9.43	9.67	9.97	7.74	8.98	29.76	30.06
Gross operating income	8.75	16.47	18.58	7.61	9.22	(1.23)	5.35	(4.09)	(26.73)
Less interest on debt	1.63	2.58	3.75	2.93	12.43	20.07	23.35	39.12	82.61
Plus other income					3.31	6.63	3.15	7.07	4.72
Less other expenses	0.28	0.90	1.86	1.87	4.79	8.70	5.93	21.63	33.39
Adjustments					0.01	(2.14)	(13.22)	(0.51)	(3.31)
Profit (loss) before tax	6.85	13.00	12.97	2.81	(4.68)	(25.51)	(34.00)	(58.29)	(141.32)
Income tax	1.36	3.07	3.63	0.79					
Net profit after tax	5.48	9.93	9.34	2.02					

Sources: Original documents of the company Luis A. Somoza D. y Cia. Ltda. (1974–77); and SSM company records (1981–85).
Note: Before the revolution, the State Sugar Mill belonged to the Somoza family. It was expropriated in July 1979.

ENABAS was aware of the indebtedness alternative available to its state-owned suppliers and took advantage of it. SSM, however, could not do the same with its own suppliers because these were mostly medium-size private cane growers. These growers were either unable to get credit to finance their accounts receivable with the same ease as the state mills or were unwilling to do so for fear that a default on their part would trigger land expropriation proceedings. As a result, SSM was concerned that a delay in paying cane growers would result, sooner or later, in a decrease in the supply of cane because growers had the option of not replanting old cane fields and of gradually replacing cane with other crops. As state sugar mills were already operating at a considerable idle plant capacity, after ENABAS failed to respond to the mills' many requests for payment, they would invariably go to the bank and get an additional loan. Eventually, this practice brought the state sugar industry to the brink of a complete financial collapse. MIDINRA had to make a formal request to the president of Nicaragua to personally intervene to redress ENABAS's systematic defaults. Meanwhile, the sugar mills saw their working capital requirements increase several times and were forced to increase their borrowing accordingly.

Second, inventories increased ninefold between 1981 and 1985, from 38 million cordobas to 348 million (Table 30). Again, this increase was associated with only a threefold increase in sales. Most of the increase in inventory was due to the fact that SSM's direct deliveries to wholesalers, which were directed by ENABAS, had to be made on consignment, since only ENABAS was authorized to purchase sugar. As a result, much of the sugar delivered by SSM remained on the books as inventory until the time that the wholesaler was authorized by ENABAS to sell it. Just as in the case of accounts receivable, the delays by ENABAS called, in turn, for additional working-capital financing.

Third, MIDINRA requested that SSM expand and modernize its mill and increase its own sugar plantation acreage as part of an overall strategy of increasing Nicaragua's sugar production.[11] New processing machinery was purchased, and additional acreage had to be irrigated and planted. These investments were responsible for most of the 257 percent increase in fixed assets between 1981 and 1985 and for additional indebtedness.

11. Interview, June 1986.

Table 30. State Sugar Mill (SSM): claims on assets (millions of cordobas)

Claims on assets	1974	1975	1976	1977	1981	1982	1983[a]	1984	1985
Current liabilities									
Accounts payable	3.16	5.96	12.68	9.14	25.66	11.29	30.99	61.02	57.26
Short-term debt[b]	4.03	17.58	16.26	10.65	126.54	210.53	235.81	425.50	933.96
Other current liabilities	2.74	7.26	13.46	11.01	5.45	7.06	7.09	15.56	40.62
Total current liabilities	9.93	30.80	42.40	30.80	157.65	228.88	273.89	502.08	1,031.84
Long-term debt[c]	1.70	10.05	10.25	19.48	10.15	26.38	58.59	63.13	68.37
Common stock	13.88	13.88	13.88	13.88	13.88	15.00	81.57	93.85	93.85
Retaining earnings	9.14	14.62	24.55	33.89	20.45	15.77	(9.74)	(43.74)	(102.30)
Earnings of last period	5.48	9.93	9.34	2.02	(4.68)	(25.51)	(34.00)	(58.29)	(141.32)
Total net worth	28.50	38.43	47.77	49.80	29.65	5.26	37.83	(8.18)	(149.50)
Total claims on assets (liabilities plus net worth)	40.13	79.28	100.43	100.08	197.45	260.53	370.31	557.03	950.71

Sources: Original documents of the company Luis A. Somoza D. y Cia. Ltda. (1974–77); and SSM company records (1981–85).
[a]1983 is the year of the first financial bailout (saneamiento).
[b]Under 18 months.
[c]Eighteen months to 5 years.

As a result of these expansions, cane production was supposed to increase by 82 percent between 1982 and 1985. Instead, in 1985 it was 15 percent lower than in 1982, mainly because irrigation was not applied regularly and most of the newly planted cane had dried out by 1985. According to SSM personnel, during that period the company was not authorized by the government to raise wages. As a result, many workers whose salary was being eroded by inflation were putting in only two-hour workdays so they could seek additional income through activities outside the company. This situation led to considerable deterioration of the cane plantations owned by the mill and to a failure to increase cane and sugar production. Idle plant capacity increased as did the need to borrow. The banks refused at first to finance these expansion projects with long-term debt; in the absence of a solid feasibility study, the mill could access only short-term loans. This was true for all state-owned sugar mills. Overall investments in the state sugar industry between 1982 and 1985 were supposed to bring its sugar output to about 4.8 million hundredweights. Production actually decreased slightly, however, and was less than half the target level. Idle plant capacity at state-owned sugar mills increased; in 1985 it was estimated at 56 percent.

Fourth, after the revolution the state company had always lost money. Between 1981 and 1985, losses increased by a factor of thirty, from about 5 million cordobas in 1981 to 141 million in 1985. Whereas before the revolution profits had been about 15 percent of sales on average, after the revolution losses averaged 25 percent of sales and were 54 percent of sales in 1985. In 1984 and 1985, SSM's net worth was negative (Table 29). Under these circumstances, the repayment of the short-term debt was not possible. The company survived only because of its unrestricted access to credit.

Just as in cotton and milk production, discussed in Chapters 3 and 4, the profitability of sugar production, which was destined in roughly equal amounts to export and domestic consumption, had been affected by foreign exchange rates and domestic price policies. The price of sugar paid to the mills was a composite of a cost-plus domestic price of sugar set by MIDINRA and an export price obtained by the Nicaraguan State Sugar Enterprise (ENAZUCAR) and converted into cordobas at the implicit exchange rate. Under pressure from state-owned mills, the government, recognizing that this price was insufficient to cover production costs, began to pay large subsidies to the mills in 1982, which were recorded by SSM as additional sales revenues (Table 31). These subsidies, however, were

Table 31. Actual calculations of the wholesale price of sugar, 1985 (cordobas per hundredweight)

1	Price to the mill[a]	1,551.92
2	Sugar subsidy[b]	877.03
3	Pro rata cost of Sugar Experimental Station	20.00
4	Pro rata cost of ENAZUCAR[c] operations	148.88
5	Pro rata cost of CANSA[d] operations	85.00
6	Sugar wholesale price[e]	2,682.83

Source: MIDINRA.

[a]This price is a composite of a "cost-plus" domestic price of sugar suggested by MIDINRA and an export price obtained by ENAZUCAR and converted into cordobas at the implicit exchange rate.

[b]Paid up to 1984 by the Ministry of Finance (MIFIN) to the sugar mill and calculated each year on the basis of the formula

$$S = \frac{(P_p - P_e) \times Q_e}{Q_d}$$

Where S = subsidy
  P_p = sugarcane price
  P_e = 1985 export price of sugar, expressed in cordobas
  Q_e = quantity of sugar exported in 1985
  Q_d = quantity of sugar for the domestic market in 1985

[c]ENAZUCAR is the state trading company that is the sole exporter of Nicaraguan sugar.
[d]CANSA is the distributor of sugar in the domestic market.
[e]Actual price paid by the wholesaler is 2,682.83 − 877.03 = 1,805.80 (cordobas).

more than offset by inflation and by the corresponding rise in the cost of goods sold, and they failed to restore the profitability of the mills. Although net sales, which after 1981 included government subsidies, had increased almost eight times between 1974 and 1985, gross profit in 1985 was about half what it had been in 1974. Interest on debt and, after 1983, administrative and other expenses contributed to the negative financial performance.

In 1984 the government decided it could no longer afford subsidies to the mills. Nevertheless, it continued to calculate sugar subsidies and to call them subsidies. After 1984, however, instead of paying the subsidy to SSM, the Ministry of Finance authorized the sugar mill to raise its price by the amount of the official subsidy. Thus, the burden of paying the subsidy was shifted from the government to the consumer.

The annual administrative procedures for setting the new price of sugar and defining the sugar subsidy were extremely complex, involving five ministries and numerous steps up to the president of Nicaragua. As a result, the mills sometimes had to wait more than six months after the milling season to know the final price of sugar. By the time the sugar mill was able to issue the final invoices for deliveries made several months before, the approved price increases had lost most of their value through inflation. The procedure also increased the amount of sugar accounted for by SSM as inventory, which gave ENABAS additional opportunities for delaying payments to SSM.

The Incentives to Borrow: New "Rules of the Game" for Credit

At the same time that SSM managers tried to deal with these pressures, two powerful incentives to borrow were developing. The first was growing negative real interest rates (Table 32). The second was a fundamental change in the structure and procedures of the bank credit boards that screened loans to state enterprises. Both incentives were initially resisted by SSM managers, who in the first three years after the revolution had not completely given up the idea of preserving the economic and financial soundness of the company. Unable to stop the indebtedness trend, five successive SSM managing directors resigned in frustration over a period of less than two years. "Eventually," explained an SSM staff member, "management had to accept the fact that the rules of the game had changed and that to have a company in a state of virtual bankruptcy no longer mattered. Once this conclusion was reached, nobody in the company had any more qualms about borrowing. Does it make sense to buy a new tractor instead of repairing the carburetor on an old one just because it looks cheaper on paper? Under the new rules of the game, it does. So we do it."[12]

One of the new "rules of the game" was that real interest rates were negative. Between 1980 and 1985, with the intent of providing an incentive to production in the farm sector, the state kept nominal interest rates on loans to the farm and agroindustrial sectors between 8 percent and 22 percent (Table 32). In addition, nominal interest rates on long-term loans were lower than those on short-term loans in order to encourage investments aimed at increasing farm output. Yet inflation was 35 percent in 1980 and reached 334 percent in 1985, according to the estimates of the Central Bank (Table 32). By other, more informal government estimates, inflation was even higher. Whatever the actual inflation rate, all data indicated

12. Interview, fall 1987.

Table 32. Interest rate structure and inflation (%)

Transaction	1980	1981	1982	1983	1984	1985
Deposits						
Savings deposits	7	8	8	8	8–10	10–12
Time deposits[a]	9	11	11	11	11–13	22–24
Loans						
Agriculture						
Short-term[b]	14	17	17	17	17	20
Long-term[c]	12	16	16	14	14	20
Livestock						
Short-term[b]	14	14	12–14	12–14	12–14	20
Long-term[c]	12	12	8–14	8–14	8–14	20
Agroindustry						
Short-term[b]	14	16	16	16	16	20
Long-term[c]	12	15	15	15	15	22
Inflation rate	35.3	33.4	22.2	32.9	50.2	334.3

Source: Central Bank of Nicaragua.
[a]One year to under 2 years.
[b]Under 18 months.
[c]Eighteen months to 5 years.

that after 1980, real interest rates on deposits were negative. To prevent a massive withdrawal of cash, the government mandated that all licensed businesses and nonprofit institutions maintain bank deposits.

Data from the Central Bank also indicate that real interest rates on loans had been negative since 1983. According to these estimates, a loan made in 1984, for example, would see its value depreciated in real terms by over 50 percent during just that year. "Under these conditions," explained an SFN executive, "the incentive to borrow is phenomenal. For anyone who has access to it, such as the state enterprises, borrowing has become a way of living."[13] Producers' revenues, however, insofar as they had been constrained by price controls, had increased at a slower pace than inflation. As a result, interest payments had often represented a real, if less obvious, cost to state enterprises.

The other "rule of the game" was that loan applications submitted by state enterprises were almost certain to be approved because of changes in the structure and procedures of the bank credit boards that screened loans to state enterprises. Since 1980, bank credit boards overseeing loan applications from state-owned enterprises had been expanded to include representatives of the particular ministry to which these enterprises belonged. These ministry representatives—many of whom had significant political credentials—often

---

13. Ibid.

became the most influential members of an otherwise technical committee. Almost invariably, the representative would support the loan applications of state enterprises. "The philosophy conveyed by these ministry representatives," explained a government official, "was that physical production was more important than money. Protecting operations that produced physical output was their goal, and this overshadowed any other consideration."[14]

At first the government felt justified in taking this position because it wanted to recover prerevolutionary output levels as soon as possible. In 1980 the banks, explained Jaime Wheelock, "essentially plugged a water hose made of money into the Ministry of Agriculture [to finance state enterprises]. Otherwise, it would have been impossible [to get production going]."[15]

After 1980, state enterprises improved their organization and control systems, but the attitude toward credit of the ministries responsible for state enterprises endured. The reactivation of the economy had created a production culture within the government that made increasing physical output the overwhelming concern of its economic branches, more important than monetary or efficiency considerations. Eventually, as macro prices penalized the financial performance of state enterprises and the setting of output goals became further divorced from cost and revenue considerations, there arose a vicious circle of ever-increasing indebtedness and continued credit extensions. If a state enterprise asked for a loan that did not seem justified to bank officers or that lacked proper documentation, the ministry representative would frequently overrule the credit board to obtain its approval. When, in some cases, the representative was unable to prevail, the minister would make a personal call to the president of the bank, and the loan was granted anyway.

An instance involving SSM is illustrative. An SSM investment project in plant expansion, irrigation, and additional cane fields that involved a loan of approximately 105 million cordobas was first turned down by the bank in 1981 because it was not documented and no feasibility study had been conducted to support it. SSM, pressured by MIDINRA, nevertheless proceeded with the project, certain as it was that eventually the ministry representative on the credit board would prevail. When the bank learned that the state enterprise had gone ahead with the expansion despite the denial of

14. Ibid.
15. Jaime Wheelock Román, *Habla la dirección de la vanguardia* (Managua: FSLN, 1981), p. 262.

credit, it felt compelled to grant a short-term loan, called "Advance on Project to Be Approved," lest it be accused of having delayed the implementation of the project and be held responsible for lost incremental output. In June 1982 the SSM feasibility study regarding the investment project, already partially implemented, was finally completed. By then, however, bank officers felt that the study was irrelevant because they knew that the loan was going to be approved anyway. Therefore, they paid little attention to the content of the study, which stated that the project was so profitable that even in the event of a substantial fall in sugar prices the company would be able to repay the loan on time. One year later, SSM was bankrupt and had to be salvaged by the government bailout. By then the bank had given up its supervisory function over its portfolio of state enterprise loans, and a demoralized staff had begun to leave the banking system. This turnover only reinforced the influence of the ministry representative, as the new bank loan officers, lacking in qualifications and experience, were unable to argue convincingly against the approval of loans.

The general phenomenon was described to me in the following terms by a prominent executive of the National Financial System:

> In 1981 and 1982, the banks were already facing massive loan defaults by the state enterprises. A typical scene would develop as follows. A ministry representative would come to a bank and ask for a loan on behalf of a state enterprise. The bank officer would ask, "What do you need it for?" The ministry representative would reply, "That's none of your business. Just give me the loan." The bank officer would then say, "If you don't tell me what you need the loan for, I am not going to give it to you." The ministry official would sometimes reply, "If you don't give it to me, I am going to denounce you as a Somoza sympathizer who is trying to undermine production." If the bank officer continued to refuse, the issue would be resolved between the minister and the president of the bank, and the loan was invariably approved. Eventually, the loan officer would seek a job elsewhere. In this way, without the government realizing it, the country lost most of whatever capable financial analysts it had. After 1985 the government began reversing its policy, but the damage had already been done, and decisions based on unsound economic analysis were perpetuated. State managers no longer had to contend with someone questioning the economic viability of their activities. That's partly why, in 1987, we ended up with more than 6,000 percent inflation."[16]

---

16. Interview, fall 1987.

Conclusion

An expansionary credit policy sustained by subsidized interest rates, along with facilitating credit decision-making mechanisms specially designed for the state enterprises, proved an irresistible vehicle for massive indebtedness for many state firms afflicted by financial problems, low labor productivity, and idle plant capacity. This massive indebtedness appeared all the more justifiable to state enterprise managers because financial difficulties and inefficiencies had become largely exogenous variables about which little could be done short of altering or contravening the government's macro price policies. The resultant administrative culture promoted production regardless of costs and revenues, removing whatever inhibitions to further indebtedness still remained among state enterprise managers. The generous credit policy acted, under the circumstances, as a temporary palliative to an impending financial crisis. Perhaps more important, it perpetuated inefficient production and further postponed the need to revise foreign exchange and domestic price policies.

A set of simultaneous pressures and incentives pushed state-owned enterprises into heavy debt. In the case of SSM and the state sugar industry, the pressures to borrow came from the need to finance growing accounts receivable, inventories, and fixed assets in the presence of increasing financial losses, as well as from negative interest rates and unrestricted access to credit. These pressures proved more powerful than did the actions that individual state managers could realistically take to preserve a relatively healthy financial structure.

The availability of easy and inexpensive credit made borrowing a much more feasible alternative than collecting payments from other state enterprises. Had the interest rate been positive, many investments would have appeared economically undesirable to the ministries that first suggested them and later asked state enterprises to implement them.

If wage rates at state-owned enterprises had been higher, laborers would not have had to seek additional part-time employment outside the firm in order to feed their families, thereby compensating for the erosion of about 70 percent of the purchasing power of their salaries between 1981 and 1985. In the case of SSM, the consensus of both management and workers was that it would have prevented the loss of new cane fields due to poor irrigation by dissatisfied workers. The incremental output resulting from these cane fields would have reduced idle mill capacity and increased revenues from sugar sales. State managers, however, found it much easier to borrow than to oppose mandatory wage policies.

An overvalued exchange rate and domestic food price policies also contributed to the negative financial results of state enterprises. In the case of sugar, managers of state-owned enterprises initially fought for better mill prices and obtained a sugar subsidy from the government. But the subsidy soon proved insufficient to compensate for low mill prices and high inflation. Borrowing to ensure the continuation of production operations was found to be a more expedient strategy than initiating a prolonged struggle over price increases, the ultimate outcome of which was uncertain.

In effect, credit provided an easy way to avoid confronting the issue. In the opinion of many government officials, credit policy managed to postpone by at least two years the need to revise macro policies. This revision came in 1985, when the national economic plan for that year finally acknowledged that the economy was out of control. The government devalued the exchange rate by 180 percent, substantially increased farm-gate prices, and more than doubled both wage rates and interest rates. Pricing mechanisms, however, were left basically unchanged. In the opinion of many managers of state-owned enterprises, the measures amounted to "too little, too late."[17]

---

17. State managers considered the inability of price formation mechanisms to adjust to ever-growing inflation the major reason why state enterprises decreased output after 1985 and joined the black market. At the end of 1987, inflation was estimated to be in excess of 7,000 percent. Primarily as a measure against inflation, in early 1988 the government eliminated the cordoba as the domestic currency and introduced the new cordoba.

7

Palliatives

> To handle the state-owned enterprises in an irresponsible way would amount to . . . transforming the July 19 [1979 Sandinista] victory over Somoza into a Pyrrhic victory.
> —Henry Ruiz, Minister of Planning, 1980

The price revisions of 1985 were a feeble attempt to correct the growing distortions in the Nicaraguan economy. The reforms failed to halt either a decline in the gross national product (GNP) or an acceleration of inflation. In 1985 GNP fell 4.1 percent and inflation reached 334 percent.[1] Within MIDINRA all but a handful of enterprises recorded losses. This poor performance of the ministry's enterprises, coupled with the sorry state of the national economy, prompted a vigorous debate within MIDINRA. The central discussion revolved around the role of the state in conditioning the behavior of the managers of the enterprises.

In 1987 the debate became public. A continued decline in GNP and an acceleration of inflation lent credence to those both within and outside the ministry that came to believe that state intervention in the economy detracted from the prudent administration of the country's resources entrusted to public administrators. The minister and vice-ministers of MIDINRA decided that the enterprises should become more autonomous from state intervention and, equally important, that administrators should be encouraged to manage the enterprises in the style of their private counterparts. There was no single directive to this effect, but the tone of the ministry began to

---

1. Figures on GNP and inflation in this chapter are from Comisión Económica para América Latina (CEPAL), *Notas sobre la economía y el desarrollo*, no. 519/520 (Santiago: CEPAL, December 1991), pp. 39, 42.

change. Accounting, for example, began to be taken more seriously. Directors and other ranking administrators were increasingly appointed on the basis of their perceived technical skills and less for their political ties to the FSLN. And banks began to request more financial data before they offered credit.

In 1987, too, MIDINRA altered the organization of its corresponding enterprises. These enterprises were formed into twelve corporations on the basis of their principal activity, such as the cultivation of sugar or cotton. Not every enterprise was included in one of the corporations, but over 90 percent were. A board of directors headed these corporations, and one of MIDINRA's vice-ministers led the board, which also included individuals outside the ministry, such as bankers.

There were various reasons for establishing the corporations. One prominent reason was to provide better organizational control. Previously there was no one line of command between the ministry and the 104 enterprises. But the central reason for the establishment of the corporations was to separate the assorted functions of the ministry from the enterprises' efforts to produce goods and services efficiently and profitably. The corporations were routinely said to be useful in helping the enterprises "professionalize" and "be more businesslike." Suggestively, the minister himself, a member of the Sandinista National Directorate, did not sit on any of the boards of directors.

Perhaps a more significant change was the ministry's instruction to the enterprises to begin ceding land to landless peasants willing to form cooperatives. Many enterprises had land that was either not productively employed or was employed in a manner marginal to the enterprises' central activities. Some enterprises, in fact, had up to 40 percent of their land fallow. Normally the directors of the enterprises selected the parcels that would be distributed to peasants under the agrarian reform. The intention was threefold: to provide needy peasants with land, to curry political favor with the rural poor majority, and to contribute to the enterprises' efficiency by making them more "compact." As a result of these allotments, the share of agricultural land held by the enterprises fell from about 23 percent in 1983[2] to 12 percent in 1988. The contribution of the enterprises to agricultural production, however—and their monopo-

---

2. MIDINRA, *Marco estratégico del desarrollo agropecuario*, vols. 1, 3 (Managua: MIDINRA, 1983).

lization of resources—continued to be disproportionately high in comparison to their share of agricultural land.

In February 1988 the government announced a set of austerity measures. These measures were decidedly more vigorous than those promulgated in 1985. There was a forceful devaluation, cuts in government spending, reductions in credit, an end to state-mandated wages, and a decline in state subsidies for a host of services and products, including foodstuffs. Perhaps most controversial was the announcement that an effort would be made to cut the state payroll by dismissing thirty thousand employees. The austerity measure provoked much political criticism that the Sandinistas were responding to the country's economic difficulties like "capitalists." But President Daniel Ortega and his fellow members of the Sandinista National Directorate increasingly felt that what was most important *politically* was economic stabilization.

A principal effect of the February measures was to lessen the overvaluation of the cordoba. MIDINRA's firms responded by attempting to employ fewer imported inputs. Those firms engaged in activities that employed considerable imported inputs suffered. And it became increasingly difficult for the enterprises to obtain credit for new investments.

The effect of the austerity measures on the economy at large was more pronounced. Economic activity declined precipitously. Indeed, in 1988 GNP contracted 13.4 percent. But the measures were incapable of halting the ever-growing breach between government expenditures and revenues. The gap was filled by what was called "inorganic emissions"—the printing of currency. Continued debasement of the cordoba was inevitable. Inflation for the year was 33,603 percent.

With inflation so high, many of MIDINRA's enterprises suddenly became "profitable." But it was widely recognized, including within the firms, that the gains were illusionary. Agricultural enterprises were likely to have the bulk of their expenses at the beginning of the agricultural season and their revenue at its end. And with inflation at over 30,000 percent, the cost of servicing debt became meaningless, as did depreciation. These "gains" from inflation were temporary, though. Beginning with the February 1988 reforms, the government all but abandoned its efforts to set macro economic policies. Workers began to demand biweekly adjustments of their wages, and the banks began to calibrate their interest rates monthly.

The macro economic environment in which managers were required to administer factories and farms became increasingly anarchical. For example, toward the end of 1988 and throughout 1989, there were wide gyrations in real interest rates. One month the real interest rate was 60 percent positive, another month it was 30 percent negative. These differences arose because interest rates were set on the basis of the anticipated inflation for the month. Predictably, there were many "poor guesses." Still, throughout the tenure of the Sandinistas, real interest rates, computed on an annual basis, were negative.

At the beginning of 1990, administrators of state enterprises, in both the agricultural and agroindustrial sectors, were managing their resources with less intervention from the ministry's central office in Managua. They were also more likely to say no to requests from other state entities, such as the army, for favors that entailed a cost. Demands, for example, to borrow trucks were likely to be met with the response, No, I have a budget and that is not included. Labor productivity began to improve. And directors became more aggressive in looking for marketing alternatives other than state monopsonies. Ventures with private enterprises became more common. The idea was even broached and discussed of putting representatives of the private sector on the board of directors of MIDINRA's twelve corporations.

What did not change in the state enterprises, though, was the continued employment of more laborers than was necessary. "Overemployment" of laborers is estimated to have been 30–40 percent throughout the tenure of the Sandinistas. An unspoken policy in rural Nicaragua was that workers could not be dismissed. What also did not change was the hellish managerial setting, characterized by widely fluctuating prices, erratic supplies, contradictory incentives, and political strife and uncertainty.

In the end, the attempts by the government to create an environment that was more conducive to the efficient and profitable use of resources failed. In real terms, the state enterprises continued to lose money. And credit to sustain their operations was never refused. Credit ceased to be unlimited, but no state enterprise was ever closed because of its inefficiency or losses. Directors were encouraged to be more efficient, to pay more attention to financial criteria in making decisions, but the environment in which they worked never demanded—or even facilitated—significant compliance.

After their electoral loss and before Violeta de Chamorro assumed the reins of government, the Sandinistas indulged in a last populist splurge. The government generously increased the salaries of public employees and otherwise committed the new government to a host of untenable expenditures. Indeed, in the last month of Sandinista rule, April 1990, the government deficit equaled the entire deficit of 1989. Intentionally or not, seeds of future economic and political strife were planted.

8

Outcomes and Conclusions

> The Nicaraguan economy is behaving like a broken jukebox: we selected a tango, but got a boogie-woogie instead.
> —Government official, Nicaraguan Ministry of Planning, 1985

Nicaragua under Sandinista rule never had a centralized planned economy. There was nothing approximating Lenin's "war communism" or Stalin's five-year plan. The planning ministry that was established in the aftermath of the Sandinista victory was intended only to coordinate the activities of other ministries. It bore no relationship to the Soviet Union's Gosplan or Cuba's Central Junta of Planning. Furthermore, the modest aims of the Nicaraguan Ministry of Planning proved elusive, and it was dissolved after three years.

The Sandinistas did attempt, however, to use the powers of the state broadly. Their adroit consolidation of control over the state enabled them to enhance the economic authority of the government apparatus, especially through nationalization and the establishment of state monopolies and monopsonies. And the Sandinistas boldly employed the traditional fiscal and monetary prerogatives of the state. By design, the instruments of government were to intervene in markets. The economy would be a "mixed economy," defined to mean that there would be coterminous private and public ownership of land and capital. But no economic actors, including those who were private, were to be immune from pervasive government initiatives and directives.

Thus, although the Sandinistas did not engage in central planning, they most surely had ambitious policies. And they had at their command both the resolve and the instruments of government necessary to ensure that their policies had a decisive impact throughout Nicaragua's economy. The nine-member leadership body, the National Directorate of the FSLN, formulated the most important policies, which once devised were, more often than not, simply issued as decrees. Ranking bureaucrats, nearly always members of the FSLN themselves, implemented the policies. It fell on these cadres to give specificity to policies and programs. The sensibleness and coherence of government policies depended in the final instance, then, on the sensibleness and coherence of the Sandinista National Directorate. The political leadership retained its governing responsibilities; it did not entrust the administration of the polity and the economy to bureaucrats. The FSLN behaved like a Leninist party in the limited but important sense that it served as the nervous system of the government.

Reviewing the entire tenure of office held by the Sandinistas from July 19, 1979, to April 15, 1990, it is clear that policy decisions were not made on the basis of any ironclad ideological or political script but instead emerged, almost piecemeal, on the basis of idelogical predispositions and political calculations of opportunities and constraints. In other words, the Sandinistas continually reviewed and revised decisions, and they did so with a calculation of how their interests would be best served. Decisions were not always made easily, but they were made consciously.

The Sandinistas assumed control of Nicaragua with the laudable commitment of ending the poverty and inequality that had characterized the country. Agriculture was acknowledged to be the necessary cornerstone of the regime's efforts at broad-based development. But the former emphasis on the generation of agroexports was tempered with an insistence on food security, nutritional well-being, and a more equitable distribution of land and income. The Sandinistas were convinced, correctly, that the redistributive component of their vision for Nicaragua would generate powerful enemies. They saw no other political group capable of defending the interests of the poor against these enemies. Thus, they concluded that the pursuit of their development goals was inseparable from their own political survival.

Upon assuming power, the Sandinista leadership was quick to issue decrees and to position the machinery of government to pursue their development goals. Perhaps the most important steps were the nationalization of the assets of Somoza and his associates and the subsequent organization of this land and capital into state enterprises. These were to be the "axis of recovery" and the "economic vanguard of the Revolution." State enterprises and those of the suspect, but tolerated, private sector were to be guided by macro economic policies formulated and promulgated by the Sandinista leadership.

It was widely suspected by all that Sandinista economic policies would not be well received by the private sector. There were marked conflicts of interests, heightened by revolutionary rhetoric and political uncertainty. This book has shown in detail, though, that even for public enterprises, Sandinista macro policies were extremely problematic. They were more than just conflictual; they were unfeasible. Sandinista economic policies created a hellish managerial environment for cadres trying to reactivate—and ultimately spur—agricultural production. The chapter-length inquiries into four critical macro policies—foreign exchange, food prices, wages, and interest rates—demonstrate that the Sandinista leadership thwarted its own noble aims. Sweeping macro policies, to which the state enterprises were bound, were at odds with the efforts of those in the fields and factories to generate wealth. The worst enemy of those toiling in the state enterprises was the state itself.

Sandinista macro economic policies did more than create managerial headaches for cadres. There were a host of unintended and undesired results: the state enterprises' persistent financial losses, their inability to service their heavy and growing debt, the underutilization of their productive capacity, and an unsatisfactory and deteriorating level of labor productivity. The enterprises absorbed a growing amount of increasingly scarce national resources with only meager returns. The cumulative cost of this deficient, and sometimes disastrous, performance affected the entire economy, contributing significantly to a fall in production and labor productivity, growing plant and equipment inefficiency, widespread speculation, and unprecedented chaos in prices. Lackluster performance meant, too, that credit was not repaid, which brought the national banking system close to bankruptcy. The debt was monetized, leading to a surge in inflation beginning in 1985. Inflation became hyperinflation, peaking at 33,000 percent.

Nicaragua's dismal economic performance under Sandinista reign cannot be traced wholly to the country's state enterprises. The counterrevolution, most visibly, took an enormous toll. And other problems were evident, the seriousness of which cannot be ignored. The argument here is only that although it was expected that the state enterprises would provide resources to combat anticipated obstacles and constraints, Sandinista macro policies derailed that expectation. State enterprises became only another problem on an already lengthy list of problems besetting the Sandinistas. And it is likely that the poor performance of the state enterprises, which became increasingly obvious, contributed to the Sandinistas' electoral defeat in the elections of February 25, 1990.

Why did the Sandinista leadership design, implement, and defend macro economic policies that were so inconsistent with the micro foundations of the Nicaraguan economy? This searching question can only elicit speculation. Some will suggest that the pressure of a relentless counterrevolution fostered problematic economic policies on the revolutionary elite. That position seems untenable. First, the origins of ill-fitted macro policies can be traced back to the beginnings of Sandinista rule, before the counterrevolution emerged. Second, and more important, there is little evidence that such prosaic policies as those directing foreign exchange or interest rates were foisted on a helpless government. The Sandinistas operated under constraints, but they took many initiatives. These initiatives were especially evident in macro economic policy formation.

There are, though, three plausible explanations for why the Sandinistas suffered from such a damaging macro-micro schism: untenable ideological predispositions, managerial shortcomings, and a combination of the two. It is worth fleshing out each of these explanations, even if they remain more suggestive than definitive.

Although macro economic policies appear technical, they are grounded in conceptualizations of the economy and of historical processes of development. A conceivable interpretation of Sandinista policy decision making is that it was rooted in a simplistic, if not outright misleading, understanding of the Nicaraguan economy. Ownership was perhaps confused with control. The prevailing critique of the Somoza regime implied that profit from land and capital was "automatic," that only the "proletariat" creates value. The contribution of management, and managers, was slighted. Similarly, attention centered on physical production and not on financial costs and revenues. There was little appreciation for the complexity of prices and markets.

Concomitantly, the conceptualization of the private sector led the Sandinistas to expend considerable effort to control perceived profits. They established state monopolies and monopsonies and instituted price controls. It is unclear to what extent these initiatives were undertaken to control the private sector and to what extent they were instituted to facilitate development. But it does seem that the perceived necessity of controlling the private sector had an impact on macro economic policy formation. Indeed, a less charitable interpretation, one frequently advanced by the Nicaraguan private sector, is that Sandinista macro policies were designed to "squeeze" the private sector and, in their destructiveness, ended up derailing the performance of state enterprises. The private sector survived only through manipulation of the black market, speculation, foot dragging, and other opportunistic strategies. While even strident opponents of the Sandinista regime do not suggest that its macro economic policies were shaped exclusively by hostility between the state and the private sector, it does seem reasonable to suggest that the government's antagonism to the private sector contributed to the formation of macro economic policies that were problematic for all producers, both public and private.

A very different interpretation as to why the Sandinista leadership promulgated macro economic policies that were detrimental to the performance of state enterprises is that the regime did not have the necessary administrative capacity to manage the economy. The establishment of such a large state sector required managerial skills that were neither available nor forthcoming. It was reasonably simple for policies to be formulated by the nine ruling *comandantes* in Managua. But their implementation throughout rural Nicaragua was slow and cumbersome. Worse, there was a decidedly inadequate ability to monitor performance and respond to problems with agility. There were also serious, if unquantifiable, problems with incentives.

The gap between soaring managerial needs and severely limited administrative capacity is easy to fathom. Despite the proliferation and burgeoning of bureaucracies, skilled and experienced administrators were scarce. Those employed did not always have skills commensurate with their responsibilities. This shortcoming was especially apparent outside of Managua. Material resources were scarce, too, occasioning obstacles and delays. Communication systems were often unreliable. Finally, political strife diverted what resources were available and distracted attention away from what must have seemed in contrast to be only mundane administrative tasks.

Conversely, the statist development model of the Sandinistas led to soaring state responsibilities. Tasks previously performed by the market, the most significant of which was price setting, now had to be performed by the state. Not only were these tasks often herculean but they were also usually unfamiliar to administrators because of the extent to which the Sandinista regime permeated the economy. Even the accounting system was reworked to make it more "revolutionary." A striking, and suggestive, example of the managerial limitations of the Sandinista regime is the anecdote recounted earlier in the book of how raising the price of sugar necessitated the involvement of five ministries, President Daniel Ortega's approval, and six months of time.

A third interpretation of why Sandinista macro economic policies were so out of synchrony with the micro foundations of the Nicaraguan economy would place the blame on a combination of untenable ideological predispositions and managerial limitations. This synthetic interpretation is perhaps the most credible. Naive or erroneous assumptions about the Nicaraguan economy may have led to the formation of problematic economic policies. These policies were difficult to implement and coordinate with any degree of finesse or agility. The lack of clarity about these difficulties may have retarded a rethinking of troublesome policies. But, in any case, ideological predispositions did not favor such a rethinking. Unfortunately, but understandably, the interplay of ideological predispositions and managerial limitations cannot be traced out systematically. At the least, it seems plausible that there was such an interactive relationship.

In the end, it can only be speculated as to why the Sandinista leadership pursued macro economic policies that worked at such cross-purposes with those who labored in factories and farms to produce goods and services. This book, though, has detailed the multitude of ways in which the Sandinistas undermined their own noble aspirations to meet the nutritional needs of Nicaragua's poor majority. Those who labored on shop floors and in fields surely made their own share of mistakes. But the data presented here show that even with diligent labor it was hard not to fall prey to a host of economic problems. Furthermore, Sandinista macro economic policies, more often than not, undermined incentives even to labor diligently.

Sandinista macro economic policies were a tragedy. An opportunity to right decades of benign neglect and outright exploitation was missed. Dreams and personal sacrifices were cruelly frustrated. And needs, including the basic nutritional needs of the poor, were left unfulfilled. Indeed, Nicaragua became decidedly poorer during the Sandinista tenure. Given the wreck wrought on postrevolutionary Nicaragua by counterrevolution and political strife, it could be suggested that the contribution of misguided economic policies to Nicaragua's misery was small. The evidence presented in this book shows that it was not. Either judgment, in any case, does not absolve the responsibility to be accorded to those who shaped and guided economic policies and programs.

The inescapable conclusion of this book is that macro economic policies must be consistent with the micro foundations of the economy. What is designed and articulated in the capital must be feasible at the farms and plants where production actually takes place. The logic of the individual laborer, manager, and the firm itself cannot be ignored. The contribution of the market must be clear. And when the state elects to intervene, the resulting administrative burden must be appreciated and must rest within the managerial abilities of the state. Macro economic policies must also be consistent with each other; they cannot work at cross-purposes. In the lexicon of this book, there must be macro-micro and macro-macro consistency.

Appendix

Major state-owned agribusiness enterprises, June 1983

Enterprise	Scope
	Reporting to MIDINRA Region I
Laura S. Olivas	Coffee and cattle (136,170 acres)
Filemón Rivera	Coffee and cattle (67,944 acres).
Oscar Turcios	Tobacco (1,706 acres)
Augusto C. Salinas	Coffee (14,079 acres
Augusto C. Sandino	Coffee processing (9 plants)
Laureano Mairena	Tobacco (10,298 acres)
César A. Molina	Construction
	Reporting to MIDINRA Region II
Ricardo M. Aviles	Cotton, rice, and tobacco (42,851 acres)
Oscar Turcios	Cattle and dairy (151,575 acres)
German Pomares	Sugarcane and 1 sugar mill (21,671 acres)
EMBANOC	Bananas (6,453 acres)
AMOLONCA	Vegetable-processing plant
AGROMEC	Farm services (306 tractors, 9 trucks)
DTSARA	Aerial fumigation services (15 planes)
ETSARA	Aerial fumigation services (33 planes)
Carlos Agüero	Cotton, rice, and sorghum (26,644 acres)
Jorge S. Bravo	Dairy and cattle (152,988 acres)
PAGRONICA	Basic grains (44,738 acres)
Arlen Siu	Ginger processing
Hilario Sánchez	Cotton and rice (18,550 acres)
	Reporting to MIDINRA Region III
Ramón Raudales	Cattle (154,074 acres)
Adolfo García	Coffee, bananas, and coffee processing (9 plants) (23,505 acres)
Roger Deshón	Dairy (17,875 acres)
Armando Talavera	Rice, corn, and bananas (28,255 acres)
Julio Buitrago	Sugarcane and 1 sugar mill (9,970 acres)

Enterprise	Scope
Ernesto Corea	Coffee and cereal processing
Gustavo Argüello	Cassava and cassava processing
AGROMEC	Farm services (machinery)
ETSARA	Aerial fumigation services
	Reporting to MIDINRA Region IV
Pikín Guerrero	Coffee and dairy (29,435 acres)
Camilo Ortega	Cattle, cotton, and basic grains (107,508 acres)
Gaspar García	Cattle (88,146 acres)
Jorge Camargo	Cattle (9,040 acres)
Comandante Exequiel	Cattle and basic grains (125,762 acres)
Benjamín Celedón	Sugarcane and sugar processing, 1 mill (13,690 acres)
Javier Guerra	Sugarcane and sugar processing, 1 mill (4,890 acres)
Claudia Chamorro	Fruit and vegetable growing and processing (1,373 acres)
Mauricio Duarte	Coffee processing (9 plants)
AGROMEC	Farm services (machinery)
Juan J. Quezada	Construction
	Reporting to MIDINRA Region V
Iván Montenegro	Cattle, dairy, coffee, and citrus fruits (118,126 acres)
Carlos Huembes	Cattle (202,949 acres)
Sergio Mendoza	Cattle (90,161 acres)
Pablo Ubeda	Cattle (66,606 acres)
Modesto Duarte	Cattle, beef, coffee, coffee processing (1 plant), and citrus fruits (104,133 acres)
EMARROZ	Sugar (10,764 acres)
	Reporting to MIDINRA Region VI
Chale Haslam	Cattle and coffee (56,434 acres)
Jacinto Hernández	Cattle, dairy, and coffee (165,226 acres)
Juan D. Muñoz	Coffee and cattle (50,597 acres)
Oscar Benavides	Basic grains, vegetables, and cattle (54,530 acres)
Jorge Vogl	Coffee and cattle (87,650 acres)
Antonio Rodríguez	Coffee processing (10 plants)
Noel Argüello	Cattle breeding (3,940 acres)
	Reporting to MIDINRA Special Zone I
Victor Ruiz	Cocoa and coconut (7,140 acres)
Peter Ferrera	Cattle and dairy (11,934 acres)
	Reporting to MIDINRA Special Zone II
Camilo Ortega	Sugarcane and sugar processing, 1 mill (3,430 acres)
Neysi Rios	Cattle (13,362 acres)
Enrique Campbell	Coconut and coconut processing (3,927 acres)
	Reporting to MIDINRA Special Zone III
Marcos Somarriba	Cattle (145,975 acres)
Hilario Sánchez	Cattle (134,861 acres)
Juan M. Loredo	Rice (6,970 acres)
AGROMEC	Farm services (machinery)

Enterprise	Scope
	Reporting to MIDINRA Central Offices
EDGRA	Services to cattle ranches
ENAMARA	Beef-packing (7 plants), pork-packing (1 plant), and meat by-products–processing (2 plants)
ENARA	Poultry and eggs (15 farms, 2 plants)
ENPRA	Animal-feed processing (2 plants) and hogs (6 farms)
ENIRA	Artificial insemination services
R. Alvarado (Chiltepe)	Dairy farm and cattle breeding
ENILAC	Milk processing (4 plants)
INFRA	Construction
NUTRIBAL	Animal-feed processing
EIERA	Research
ENARA	Auditing
PROCOMER	Procurement, storage, and distribution of fruits and vegetables
TECHNOPLAN	Engineering and project designing
CACE	Computer data processing services
EMPROSEM	Production, procurement, and distribution of certified seeds
AGROINRA	Supervises state agroindustrial enterprises (TIMAL, SOVIPE, ENILIB, EMFARA, MacGregor, Valle de Sebaco, Palma Africana, Agroindustrial del Caucho)
SUMAGRO	Import and distribution of farm inputs
ENIA	Sole importer of most fertilizers, pesticides, and other agrichemicals
PROAGRO	Producer and distributor of agrichemicals
TRANSAGRO	Transportation of farm products and inputs
ENTRA	Repairs and maintenance of farm equipment
ECAMI	Repairs and maintenance of communications equipment
AGROMAQ	Sales of farm equipment
	Reporting to the Ministry of Foreign Trade (MICE)
ENCAR	Sole exporter of Nicaraguan beef
ENAL	Sole exporter of Nicaraguan cotton
BANANIC	Sole exporter of Nicaraguan bananas
ENCAFE	Sole exporter of Nicaraguan coffee
ENAZUCAR	Sole exporter of Nicaraguan sugar
ENIPREX	Exporter of nontraditional products (fruits, vegetables, ginger, tobacco)
	Reporting to the Ministry of Internal Trade (MICOIN)
ENABAS	Procurement, storage, and distribution of basic grains
	Reporting to the Ministry of Industry (MIND)
COIP	Holding company controlling state-owned industrial enterprises, some of which process farm products

Index

agrarian policy. *See* food policy, Nicaraguan
agribusiness systems approach, 6
AGROINRA, 130
AGROMAQ, 168
AGROMEC, 46, 67
APP (Area of People's Property). *See also individual enterprises*
　debts of, 61, 174–175, 179, 190–191
　financial controls, 12, 62, 177, 182–188
　formation of, 2, 40, 43, 66–68, 141
　goals of, 68–70, 147
　performance of, 2–3, 8, 52–58, 61–63, 70, 94, 178–179, 195
　as policy instrument, 9, 45–46, 59–60, 191, 195–197
ATC (Association of Rural Workers), 68, 140
Austin, James, 15, 35, 46

banana production, 23
BANANIC (Nicaraguan Banana Enterprise), 46
Banco de America (BAMER), 46, 174
BANIC (Nicaraguan Bank), 46, 174
banks, 45, 173–174, 191–192
　foreign, 174
　relationships with APP, 74–75
"basic needs" approach, 5, 8
BCN. *See* Central Bank of Nicaragua
BCP (Popular Credit Bank), 174
beef commodity system, 105–122, 124–130, 133–137
beef consumption, domestic, 114
beef production, 10, 17, 21–22, 26–27, 106, 112. *See also* cattle ranchers
　expansion of, 38–39, 105–106

Biderman, Jaime, 27
BINMO (Banco Inmobilario), 174
Biondi-Morra, Brizio, 12–13
black market, 11, 74, 79, 122–127, 133–136, 162–171
BND (National Development Bank), 46, 174

cacao production, 21–22
Carrión, Luis, 24
case method, 13
cattle ranchers, 57, 107, 122–124
CDD (Certificate of Exchange Availability), 82–86, 89, 95, 98
Central American Common Market (CACM), 52
Central Bank of Nicaragua (BCN), 45, 68, 131, 173–174
　analysis of state enterprises, 85
　commissions, 82
　economic indicators, 160–162, 189–190
　relations with state enterprises, 85, 182
Central Sandinista de Trabajadores (Sandinista Workers Center), 148–150
Certificate of Exchange Availability (CDD), 82–86, 89, 95, 98
Chamorro, Violeta de, 198
Chinandega, 25
CIERA (Research Center for Economics and Land Reform), 24, 95
　economic indicators, 162–163
coffee production, 17, 22–23
Colburn, Forrest, 139
commodity modeling, 6
Common Market. *See* Central American Common Market

211

"conservatives," 22
consumer purchasing power, 37, 39, 128, 135, 160–163, 169–171
contras, 4, 52, 154, 163, 202, 205
CORFIN (Financial Corporation of Nicaragua), 46, 67, 174
"cost-plus" pricing system, 39, 57, 79–88. See also pricing system, agricultural
cotton commodity system, 65–69
  performance of, 70, 73
cotton growers, 57, 74, 83–84, 86
cotton pricing system, 10, 68, 76–94. See also "cost-plus" pricing system
  "pool pricing," 77–79
cotton production, 10, 17, 23–26, 63–98
  area planted, 64–65, 69–74
  role in food system, 64–65, 70, 97
  yields of, 64, 70–75
cottonseed oil, 27, 67–68
counterrevolutionary war, 4, 52, 154, 163, 202, 205
Credit and Services Cooperatives (CCS), 43
credit policy, 11, 39, 174–194, 198
CST (Sandinista Workers Center), 148–150

decision-making process, 130–133, 137, 188–189, 191
Deere, Carmen Diana, 146
devaluation, 76, 88, 94. See also exchange rates
development models, 8
development plan of 1952, 17
DOGE (Entrepreneurial Organization and Administration Division of MIDINRA), 62

economic policy, 100–101, 191, 201–203. See also food policy, Nicaraguan
Economic Reactivation Plan (1980), 61
economy, Nicaraguan
  colonial, 19, 21–22
  postrevolution, 47–48, 195, 197–198
  prerevolution, 17, 28–29
elections (1990), 11
employment, 163–164, 198. See also workers
EMPROSEM, 46
ENABAS (National Supply Company), 45–46, 183–185, 189
ENAL (Nicaraguan Cotton Enterprise), 46, 67–68, 74, 77–78, 81
ENAMARA (National Enterprise of Land Reform Slaughterhouses), 105, 109–113, 115–122, 124, 126, 129–130, 135–136

ENAZUCAR (Nicaraguan State Sugar Enterprise), 46, 187
ENCAFE (Nicaraguan Coffee Enterprise), 46
ENCAR (Beef Enterprise), 111, 120–122
ENIA (Nicaraguan Enterprise of Farm Inputs), 66–67
ENILAC (Enterprise of the Milk Industry), 105, 111–113, 115–118, 123–126, 128, 131, 135
Esteva, Gustavo, 15
ETSA, 46, 67
European Economic Community (EEC), donations from, 113
exchange rates, 88–90
  and price policy, 76, 79–81, 84–94
export prices, 73, 97
exports, 48, 105–106
  reductions in, 97, 113–114, 118–122
export sector, 17–19, 57–58, 63–66, 75
  contribution to economy, 35, 103
  historical development, 21–27

FAGANIC (Nicaraguan Federation of Cattle Rancher Associations), 106, 122, 124
Falcon, Walter, 7–9
FNI (National Investment Fund), 46, 130
FONDILAC (association of private milk producers and processors), 122–124
food policy, Nicaraguan. See also pricing system, agricultural
  failure of, 3–4, 10
  impact on revolution, 14–15
  impact on state enterprises, 96
  formulation of, 35–38, 101–103
  goals of, 59–62, 151, 175
  premises of, 15–17, 31–32, 42, 59–60, 156
  prerevolution, 29, 102
"food price dilemma," 6
food security, 31–32, 38, 151
food system, Nicaraguan
  growth capability, 58, 103
  historical stages, 19–27
  performance of, 46–55, 99–100, 103–105, 152
  prerevolution, 30, 101–102
  role in economy, 17–18, 103
Ford Foundation, 12
foreign debt, 35, 59
foreign exchange, 10, 34–35, 52, 64–65
  availability of, 73
  drop in, 57–58
  generation of, 86–87
Fox, Jonathan, 35, 46
FSLN. See Sandinista Front for National Liberation

General Cattle Division of MIDINRA (GCD), 131
General Economic Division of MIDINRA (GED), 131
gold production, 23
Granada, 22

"historical vacation" (*vacación histórica*), 141-142, 150
hyperinflation, 10-11, 48, 189-190, 195, 197

import dependence, 30, 34-35, 113, 118, 152
INCAE (Central American Institute of Business Administration), 12-13, 15, 115-117
indigo production, 21-22
INEC (Nicaraguan Institute of Statistics and Census), 126-127
inflation, 10-11, 48, 189-190, 195, 197
INFONAC (Nicaraguan Institute of National Development), 106
INRA (Nicaraguan Institute of Agrarian Reform), 70
interest rates, 11, 189-190, 198
international prices, 4-5
investment policy, 39-40

Krüger, Walter, 46

labor productivity, 47, 142-151, 163-167, 170-172
land dispossession, 21-23, 25
landowners
  large, 16, 21, 42-43
  small and medium, 43
land reform, 117, 140-142, 196
latifundios, 16, 21
León, 22, 25
"liberals," 22
livestock. *See* beef production
loans to state agencies, 52, 174-179, 182-194
local governments, 131

macroeconomists, 6
macro-micro policy consistency, 3-4, 7, 12, 203-206
Managua, 135, 168
Masaya, 25
MICE. *See* Ministry of Foreign Trade
MICOIN. *See* Ministry of Internal Trade
MIDINRA (Ministry of Agriculture and Land Reform)
  analysis of food system, 30-31, 81-82, 95, 114-117, 127, 129-130,
  134-135, 143, 146-147, 149, 179
  analysis of industrial sector, 101-105
  commissions, representation on, 82
  discrepancies in data, 127
  management of state enterprises, 111-112, 115, 130-133, 158, 185, 191, 195-198
  projections, 32, 43, 178
  relations with other ministries, 63
  relations with producers, 86
  strategy, 176
MIDINRA/INCAE research and training project, 12-13, 15, 115-117
milk commodity system, 105, 108-118, 122-131, 136-137
milk pricing system, 111, 123-128, 130-134
milk-processing plants, 10, 108-111
MIND (Ministry of Industry), 130, 132
minifundios, 16, 21
Ministry of Agriculture and Land Reform. *See* MIDINRA
Ministry of Foreign Trade (MICE), 46, 111, 130-132
  commissions, 82
Ministry of Health (MISAL), 32
Ministry of Industry (MIND), 130, 132
Ministry of Internal Trade (MICOIN), 130-133, 183
  economic indicators, 162
Ministry of Labor (MITRAB), 32
Ministry of Planning (MIPLAN), 14, 32, 62, 130, 178, 200
  commissions, 82
  studies by, 134-135
MITRAB (Ministry of Labor), 32

National Economic Plan (1985), 47, 95, 150
National Financial System (SFN), 46-47, 63, 135, 174, 179
National Food Program. *See* PAN
National Planning Council, 131
Nestlé, 108
"New Economy," 143-146, 171, 177
Nicaragua
  development potential, 17
  independence (1821), 21
  physical characteristics, 19-20
  population, 19, 21, 48
Nicaraguan Agrarian Reform Law (1981), 68
Nuevo Diario, 169
Núñez, Orlando, 24

Ortega, Daniel, 205

PAN (National Food Program), 38, 46, 130
Pearson, Scott, 7-9
peasants, 139-142, 196
  migration of, 25, 27, 135, 168

Pizarro, Roberto, 94
Plan for Livestock Development, 38, 46
Plan for Sugar Development, 39, 46
Plan for Vegetable Development, 39, 46
Plan for Vegetable Oils Development, 39, 46
price incentives, 39, 122–127, 133, 136, 168
pricing system, agricultural, 47, 57, 100, 120–122, 130–136, 187–189. *See also* "cost-plus" pricing system; cotton pricing system; milk pricing system
prerevolution, 109
PROAGRO (National Enterprise of Farm Products), 46, 67
"production incentives" approach, 5, 8

research methodology, 13
Revolutionary War (1979), 138
  roots of, 18, 21
rice production, 154–156, 158, 160, 166. *See also* State Rice Farm
rubber production, 23
Ruiz, Henry, 142

Sandinista Agricultural Cooperatives (CAS), 43, 68
Sandinista Front for National Liberation (FSLN), 139–140, 142
  development strategy, 1–2, 31, 101, 143–146, 173, 197, 200–202
  electoral defeat, 11, 198–199, 203
  labor relations, 147–148
saneamiento financiero (financial bailout). *See* APP: debts of; loans to state agencies
SFN (National Financial System), 46–47, 63, 135, 174, 179
shortages, 52, 58, 114
slaughterhouses, municipal, 107, 109–113, 121. *See also* beef production
SNOTS (National Organizing System of Labor and Wages), 157, 169
"social wage" policy, 143, 156–157, 171
Somoza, Anastasio, 1
Somoza family, confiscation of assets, 2, 40–41, 56, 67, 111, 141–143, 158, 177–178, 180, 201–202
sorghum production, as alternative to cotton, 96–97

State Cotton Farm (SCF), 10, 70–75, 88–93, 96–97
state enterprises. *See* APP
State Rice Farm (SRF), 157–169
State Sugar Mill (SSM), 11, 180–189, 191–194
subsidies, 39, 52, 113, 187–189, 194
  implicit, 90–94
subsistence farming, 19
SUCA (System of Uniform Administrative Control), 180
sugar sector, 11, 17–18, 39, 182–189
SUMAGRO, 46, 130

technology policy, 40
TECNOPLAN, 130
TEXNICSA, 67
timber production, 23
Timmer, Peter, 4–8

United Nations
  Food and Agriculture Organization(FAO), 101–105, 129
  study of Nicaraguan food system, 170–171
  World Food Program (UNWFP), 113
United States
  Agency for International Development (USAID), 29
  Department of Agriculture (USDA), 106–107
  import quotas, 107, 182
  intervention, 22
  trade embargo, 52, 113, 182

vacación histórica (historical vacation), 141–142, 150
vegetable production, 39

wage rates, 11, 143, 156–172
war. *See* counterrevolutionary war; Revolutionary War
Wheelock Román, Jaime, 22, 24, 30–31, 100–102, 191
workers, 162–168, 170–171
World Bank, 17, 24
World Food Conference (1974), 6, 9

Zelaya, José Santos, 22

Food Systems and Agrarian Change

Edited by Frederick H. Buttel, Billie R. DeWalt,
and Per Pinstrup-Andersen

Hungry Dreams: The Failure of Food Policy in Revolutionary Nicaragua, 1979–1990
by Brizio N. Biondi-Morra

Research and Productivity in Asian Agriculture
by Robert E. Evenson and Carl E. Pray

The Politics of Food in Mexico: State Power and Social Mobilization
by Jonathan Fox

Searching for Rural Development: Labor Migration and Employment in Rural Mexico
by Merilee S. Grindle

Structural Change and Small-Farm Agriculture in Northwest Portugal
by Eric Monke et al.

Diversity, Farmer Knowledge, and Sustainability
edited by Joyce Lewinger Moock and Robert E. Rhoades

Networking in International Agricultural Research
by Donald L. Plucknett, Nigel J. H. Smith, and Selcuk Ozgediz

The New Economics of India's Green Revolution: Income and Employment Diffusion in Uttar Pradesh
by Rita Sharma and Thomas T. Poleman

Agriculture and the State: Growth, Employment, and Poverty in Developing Countries
edited by C. Peter Timmer

Transforming Agriculture in Taiwan: The Experience of the Joint Commission on Rural Reconstruction
by Joseph A. Yager